# Statistical Learning
# and Data Science

Chapman & Hall/CRC
# Computer Science and Data Analysis Series

The interface between the computer and statistical sciences is increasing, as each discipline seeks to harness the power and resources of the other. This series aims to foster the integration between the computer sciences and statistical, numerical, and probabilistic methods by publishing a broad range of reference works, textbooks, and handbooks.

**SERIES EDITORS**
David Blei, Princeton University
David Madigan, Rutgers University
Marina Meila, University of Washington
Fionn Murtagh, Royal Holloway, University of London

Proposals for the series should be sent directly to one of the series editors above, or submitted to:

**Chapman & Hall/CRC**
4th Floor, Albert House
1-4 Singer Street
London EC2A 4BQ
UK

## Published Titles

Bayesian Artificial Intelligence, Second Edition
*Kevin B. Korb and Ann E. Nicholson*

Clustering for Data Mining: A Data Recovery Approach
*Boris Mirkin*

Computational Statistics Handbook with MATLAB®, Second Edition
*Wendy L. Martinez and Angel R. Martinez*

Correspondence Analysis and Data Coding with Java and R
*Fionn Murtagh*

Design and Modeling for Computer Experiments
*Kai-Tai Fang, Runze Li, and Agus Sudjianto*

Exploratory Data Analysis with MATLAB®, Second Edition
*Wendy L. Martinez, Angel R. Martinez, and Jeffrey L. Solka*

Exploratory Multivariate Analysis by Example Using R
*François Husson, Sébastien Lê, and Jérôme Pagès*

Introduction to Data Technologies
*Paul Murrell*

Introduction to Machine Learning and Bioinformatics
*Sushmita Mitra, Sujay Datta, Theodore Perkins, and George Michailidis*

Microarray Image Analysis: An Algorithmic Approach
*Karl Fraser, Zidong Wang, and Xiaohui Liu*

Pattern Recognition Algorithms for Data Mining
*Sankar K. Pal and Pabitra Mitra*

R Graphics
*Paul Murrell*

R Programming for Bioinformatics
*Robert Gentleman*

Semisupervised Learning for Computational Linguistics
*Steven Abney*

Statistical Computing with R
*Maria L. Rizzo*

Statistical Learning and Data Science
*Mireille Gettler Summa, Léon Bottou, Bernard Goldfarb, Fionn Murtagh, Catherine Pardoux, and Myriam Touati*

Computer Science and Data Analysis Series

# Statistical Learning and Data Science

Edited by

Mireille Gettler Summa, Léon Bottou,
Bernard Goldfarb, Fionn Murtagh,
Catherine Pardoux, and Myriam Touati

CRC Press
Taylor & Francis Group
Boca Raton London New York

CRC Press is an imprint of the
Taylor & Francis Group, an **informa** business

A CHAPMAN & HALL BOOK

CRC Press
Taylor & Francis Group
6000 Broken Sound Parkway NW, Suite 300
Boca Raton, FL 33487-2742

First issued in paperback 2019

ISBN-13: 978-1-4398-6763-1 (hbk)
ISBN-13: 978-0-367-38189-9 (pbk)

---

### Library of Congress Cataloging-in-Publication Data

---

Statistical learning and data science / [edited by] Mireille Gettler Summa ... [et al.].
    p. cm. -- (Chapman & Hall/CRC computer science & data analysis)
  Includes bibliographical references and index.
  ISBN 978-1-4398-6763-1 (hardback)
  1. Machine learning--Statistical methods. 2. Data mining. I. Summa, Mireille Gettler.

Q325.5.S73 2012
006.3'1--dc23
                                                2011036558

---

Visit the Taylor & Francis Web site at
http://www.taylorandfrancis.com

and the CRC Press Web site at
http://www.crcpress.com

# Contents

# *Preface*

## Introduction

Machine learning, statistical learning, computer learning, data analysis, data mining, and related data sciences often face the same issues but these domains have had quite separate historical development. This book aims at presenting recent results which have been established by some of the main founders of these domains. It gives also a first opportunity to gather different approaches in crucial, up-to-date contexts: social networks, web mining, data streams, texts, high-tech data, and other areas.

## Part I: Statistical and Machine Learning

The first five chapters are dedicated to learning and the chapters are organized from applications through to a major theoretical result in the domain.

The first chapter is about large scale social networks that contain very useful data in a very peculiar form. Chapus, Fogelman-Soulié, Marcadé, and Sauvage give an overview of exploratory and predictive methods for such data. They show how a thoughtful exploitation of social networks leads to significant improvements in telecommunications, fraud detection, and retail applications.

Many machine learning problems deal with abundant training data and limited training time. Bottou describes in the second chapter the tradeoffs associated with such large-scale learning systems. Because he involves the computational complexity of the underlying optimization algorithm in nontrivial ways, unlikely optimization algorithms such as stochastic gradient descent show amazing performance for large-scale problems.

In the third chapter, learning with privileged information (cf. Vapnik & Vashist 2009 [304]) takes advantage of supplementary information available only during training time in a manner reminiscent of a student taking advantage of the explanations of a teacher. Pechyony and Vapnik describe fast algorithms to train Support Vector Machines with the added benefits of privileged information. These algorithms work orders of magnitude faster than heretofore and therefore enable much more ambitious applicative research.

Given a set of independent and identically distributed (i.i.d.) labeled examples $(x_1, y_1), \ldots, (x_n, y_n)$, and a previously unseen pattern $x$, conformal predictors (see Vovk et al. 2005 [308]) try all possible predictions $y$ and pick the one that makes the set $(x_1, y_1), \ldots, (x_n, y_n)$ as compatible as possible with the i.i.d. assumption. In the fourth chapter Adamskiy, Nouretdinov, and Gammerman propose an extension of conformal predictors to semisupervised settings where one has both labeled and unlabeled examples.

The Vapnik-Chervonenkis statistical learning theory hinges on the uniform convergence of empirical frequencies to their probabilities. An essential theorem states that this uniform convergence occurs if and only if the VC-entropy per element converges to zero. Chervonenkis proposes in the fifth chapter a short proof that elucidates what happens when the uniform convergence does not take place, that is, when the entropy per example converges to $c > 0$. It is then possible to identify a subspace of probability $c > 0$ on which the learning machine is nonfalsifiable (the principle expanded on by philosopher Karl Popper). This result has been mentioned without proof (in Vapnik, 1998, theorem 3.6.). To our knowledge, this is the first publication of the proof in English.

---

# Part II: Data Science, Foundations, and Applications

The second part of the book includes chapters relating to the two main French founders on the one hand of data analysis and, on the other hand, of one of its principal domains of applications in social science.

## Benzécri: The Continuous and the Discrete in Data Analysis

Daniel Bennequin and Fionn Murtagh consider the themes elaborated on in chapter 6 in the following.

### The Continuous and the Discrete

We will skip discussions in antiquity of the continuous and the discontinuous (Democritus, Epicurus, and others who were proponents of atomism and the Stoic counterargument on the continuous). It has been mostly atomism that has prevailed since Boltzmann. This does not prevent mechanics, field theory, and other domains being presented nowadays based on a continuous vector field, from which discontinuities are deduced and vice versa. Hence, one can say that modern-day thinkers who seek to start with the discrete still have work on their hands as they pursue that vantage point.

Perhaps, all the same, it is necessary to cite Maxwell's cogwheel model. Consider challenges of mobility in Maxwell that Gilles Chatelet laid out very well in his book, *Figuring Space* [68].

This point of view of a discrete origin of space, of time, and of fields is adopted between the lines – by one of the inventors of strings and superstrings (that could viewed as a paradox), A.M. Polyakov [258]. See the first pages of his book Gauge Fields Strings on "excitations of some complicated medium." Even if it is especially in verbally communicated presentations that Polyakov takes up this question of the discrete origin, he presents Yang-Mills on a network, or strings as random surfaces with polymers that flatten themselves down. Furthermore, he sees renormalization as a continuous limit of a network, the details of which are unknown, and what is thereby described is nothing other than continuous "effective theories."

Starting with the discrete is also, however, what underlies the approach of Wilson, Nobel Prize winner for his theories on networks. In fact, he and other researchers (Polyakov, Kadanoff, 't Hooft) get their inspiration from the fundamental (and mysterious) correspondence between statistical systems (temperature) and quantum field theories (with time).

Finally in the same direction we should cite (even if it seems often to be in conflict with such a perspective) the work of "quantum gravity" and in particular Ashtekar and Rovelli (e.g., [8]). We may also refer to the work of Ambjørn on causal dynamical triangulation, relating to the origin of a fourth dimension.

One sees therefore that in his chapter J.P. Benzécri is not alone, even if his approach is an original one.

## Between Physics and Computer Science

Arising out of computer science rather than physics, Wolfram's *A New Kind of Science* [316] seeks to provide a computational theory of the universe through cellular automata. Patterns of all manner grow through having rules, or algorithms. J.P. Benzécri's grains and spikes are in a way reminiscent of such a computational or algorithmic framework. The interface between physics and computation has become quite diffuse. As Nielsen and Chuang [243] remark, "Quantum computation and quantum information has taught us to think physically about computation ... we can also learn to think computationally about physics." Even more generally, computational models and algorithms for the modeling of physical or chemical – or social or artistic or whatever – processes are discrete just because they are implemented on digital and not analog computers. In Benzécri's chapter, discrete origins are laid out in terms of preorders. The heart of the presentation in his chapter comes later. This culmination of his presentation is how he derives a continuous space from finite sets, and more particularly from the correspondence between such finite sets.

## Between Benzécri and Bourdieu

Chapter 7 is linked to the first one considering the common presence of J.P. Benzécri and P. Bourdieu at the Ecole Normale Supérieure of Paris in the same years (1950). Brigitte Le Roux presents, in the following, the general framework, Geometric Data Analysis (GDA), in which the recent advances of data science applied to sociology are imbedded.

Since the very beginnings GDA has been applied to a large range of domains, such as medicine, lexicology, market research, satisfaction surveys, econometrics, and social sciences. In this latter domain, the work of the French social scientist Pierre Bourdieu is exemplary of the "elective affinities" between the spatial conception of social space and GDA representations (Rouanet et al. 2000 [277]). These affinities led Bourdieu and his school to use Correspondence Analysis (CA) – especially Multiple Correspondence Analysis – consistently in the 1970s, 1980s, and 1990s (Bourdieu & De Saint-Martin 1976, 1979, 1999 [49, 42, 45]). It is commonplace to view CA as "Bourdieu's statististical method." This is rightly so, since Bourdieu's work has provided exemplary use of CA. In fact, beyond CA, a constant concern for Bourdieu was to substantiate his theories by statistical data and analyses. This essential aspect of Bourdieu's work has not been covered in detail so far by those who have written about Bourdieu's theories.

In Chapter 7, Lebaron deals with this particular domain of application of GDA techniques. He provides some landmarks concerning Bourdieu and statistics, from traditional statistics to GDA. Then he highlights the powerful link between geometry and the construction of social space – "Social reality is multidimensional" – as well as the investigation of the notion of "field" in Bourdieu's theory. On the basis of several examples, he shows the relevance of these techniques to discover the structures of social spaces.

Bourdieu passed away in 2002, but the interest of sociologist researchers in the statistical methods that he had used, especially GDA, is increasing more and more, worldwide, as evidenced too in the recent work, (Le Roux & Rouanet 2010 [197]).

## Applications

The second subsection of the second part of the book is devoted to the presentation of five methodologies, illustrated in a wide area of applications (semantics, credit risk, energy production, genomics, and ecology).

In Chapter 8 (Murtagh, Ganz, and Reddington), a framework for the analysis of semantics is developed through Correspondence Analysis, separating the context (the collection of all interrelationships), and the hierarchy tracks anomaly. There are numerous potential applications like business strategy and planning, education and training, science technology and economic development policy, although the methods are illustrated here with the semantics of a film script.

In Chapter 9 (Hand), the measures of accuracy in classification are reanalyzed with a new look at a widely used measure, the so-called Area Under the ROC Curve (AUC index). Using mathematical reassessments, the author points out the source of its weaknesses and proposes a new performance measure, labeled H.

In Chapter 10 (Da Silva, Lechevallier, and Seraoui) the special features of time-changing data streams are investigated. A clustering approach is presented for monitoring numerous time-changing data streams, and illustrated by real data from electric power production.

Chapter 11 (Mariadassou and Bar-Hen) deals with the reconstruction and the validation of phylogenetic trees with a special focus on robustness. Their approach is based on several means: bootstrap or jackknife, influence function, and taxon influence index.

In Chapter 12 (Demeyer, Foulley, Fischer, and Saporta) a Bayesian approach to Structural Equation Models (SEM) is proposed, in which the use of posterior draws allows us to model expert knowledge. Particular attention is also given to identifiability for which they use a parameter expansion.

Throughout all these chapters, the reader may appreciate the strength of the concept of Data Science, as well as the considerable extent of its applications.

## Part III: Complex Data

The last set of chapters is about complex data, that is, data which are not merely values (e.g., 32, 45), or levels (e.g., third) or categories (blue). Statistical methods and results depend most of the time on the nature of data: for example the computation of the average of two words does not make sense. More and more complex data have become available for processing, over the history of multidimensional data analysis. Since the early 20th century, K. Pearson handled multivariate numerical data tables. Psychometrics yielded boosted results for ordinal data. Major results for categorical data were established in the early 1960s in France. Pattern recognition, textual analysis, and multidimensional time series gave the opportunity for facing more complex data that were put in various frameworks such as functional data analysis or symbolic data analysis. Complex data are nowadays crucial: multimedia data, mixture of figures, images, texts, temporal or streaming data constitute the databases of companies or public statistical depositories. Part III of the book presents some approaches for time series, symbolic data, and functional data.

In Chapter 13, Gettler Summa, Goldfarb, and Vichi propose a tandem cluster analysis in order to obtain a partition on a set of multiple and multidimensional numerical time series. A principal component analysis is applied on the three-way table and an innovative weighted dissimilarity is defined for the clustering step that takes into account velocity and acceleration. A major point is that trajectories clustering is a three-way extension of usual clustering approaches; it is therefore a non-polynomial (NP)-hard problem and no guarantee to find the optimal solution is possible. Nevertheless, a multi-start procedure increases the chance to find an optimal solution through a sequential quadratic algorithm. The approach is used to find patterns among the cancer standardized death ratios of 51 countries in over 21 years.

There is a great variety of algorithms for constructing prediction trees from data, the most popular of which is CART, the acronym for Classification and Regression Trees. All such algorithms share a central idea: at each step of a recursive procedure, a subsample of the data is partitioned according to the value of a predictor, the actual split being chosen so as to maximize a measure of split quality. In Chapter 14, Ciampi proposes to radically transform the recursive partition algorithm, which is inherently local, by making it global, i.e., by using all data, at each step of the construction. Ciampi extends then the main idea of his work to a certain development of symbolic data analysis. The aim of the symbolic approach in data analysis is to extend problems, methods, and algorithms used on classical data to more complex data called "symbolic," which are well adapted to represent classes or categories. Ciampi introduces the notion of soft node, and tree with soft nodes respectively. Two examples of real data analysis are presented and the author finally summarizes the results of some comparative empirical evaluation of prediction trees with soft nodes.

Concepts in the symbolic data analysis framework are defined by intent and an extent which satisfy the intent. Symbolic data are used in order to model concepts by the so called "symbolic objects." In Chapter 15, Touati, Djedour, and Diday define three kinds of symbolic objects: events, assertions, and synthesis objects. They focus on synthesis objects: they consider several objects characterized by some properties and background knowledge expressed by taxonomies, similarity measures, and affinities between some values. With all such data, the idea is to discover a set of classes, which are described by modal synthesis objects, built up by assertions of great affinity. The authors use a symbolic clustering algorithm that provides a solution allowing for convergence towards a local optimum for a given criterion and a generalization algorithm to build modal synthesis objects. Finally they give an example to illustrate this method: the advising of the University of Algiers students.

The progress of electronic devices like sensors allows us to collect data over finer and finer grids. These high-tech data introduce some continuum that leads statisticians to consider statistical units as functions, surfaces, or more generally as a mathematical object. In Chapter 16, Delso, Ferraty, and Martínez Calvo introduce the generic terminology for this new kind of data, Functional Data (FD) and the statistical methodologies using FD are usually called Functional Data Analysis (FDA). The aim of their contribution is to provide a survey voluntarily oriented towards the interactions between FDA and various fields of the sciences whereas methodological aspects may be found in Chapter 17, Gonzàlez Manteiga and Vieu. The latter chapter focuses on the most recent advances instead of a general survey. The contributions give several examples of FD and describe the various statistical problematic involving FD: functional principal component analysis (FPCA), various distributional descriptors (mainly dealing with centrality notions), classification, and explanatory methods in FDA such as predicting scalar response from functional covariates.

# Acknowledgments

A number of young colleagues played a major role in the project that has resulted in this book:

- Solène Bienaise, CEREMADE, Université Paris-Dauphine, France

- Mohamed Chérif Rahal, CEREMADE, Université Paris-Dauphine, France

- Cristina Tortora, CEREMADE, Université Paris-Dauphine, France and Università Federico II, Naples, Italy.

We are also indebted to our colleagues

- Daniel Bennequin, Université Paris 7, UFR de Mathématiques, 175 rue du Chevaleret, 75013 Paris, France. *bennequin@math.jussieu.fr*

and in particular for contributions throughout this work,

- Brigitte Le Roux, MAP5 Université Paris Descartes, 45 rue des Saints Pères, 75006 Paris, France, & CEVIPOF/CNRS, 98 rue de l'Université, 75007, Paris, France. *Brigitte.LeRoux@mi.parisdescartes.fr*

Finally, special acknowledgments are noted by Mireille Gettler Summa and Myriam Touati to their respective spouses and children, Pierangelo and Robin Summa, Joël and Sarah Touati, for their patience and their support throughout the editing process of this book.

# Contributors

**Dmitry Adamskiy**
Computer Learning Research Centre,
Royal Holloway, University of London,
England
adamskiy@cs.rhul.ac.uk

**Avner Bar-Hen**
Université Paris Descartes, Paris 5, MAP5
45 rue des Saints Pères, 75270 Paris Cedex
06, France
Avner.Bar-Hen@mi.parisdescartes.fr

**Jean-Paul Benzécri**
33 rue de la République,
F-45000 Orléans, France
jpfbnz@wanadoo.fr

**Léon Bottou**
Microsoft, Redmond, WA, USA
leon@bottou.org

**Benjamin Chapus**
KXEN, 25 quai Galliéni
92158 Suresnes Cedex, France
benjamin.chapus@kxen.com

**Alexey Chervonenkis**
Institute of Control Science, Lab. 38,
Profsoyusnaya 65, Moscow 117997, Russia
chervnks@ipu.ru

**Antonio Ciampi**
McGill University, Montréal,
Québec, Canada
antonio.ciampi@mcgill.ca

**Alzennyr Da Silva**
BILab (Telecom ParisTech and EDF R&D
Commom Laboratory)
INFRES-Telecom ParisTech,
46 rue Barrault, 75013 Paris, France
alzennyr.gomes@telecom-paristech.fr

**Laurent Delsol**
Université d'Orléans, France
laurent.delsol@univ-orleans.fr

**Séverine Demeyer**
Laboratoire National de Métrologie et d'Essais,
LNE, 1 rue Gaston Boissier, Paris, France
sev.demey@yahoo.fr

**Edwin Diday**
Université Paris-Dauphine
Place du Mal. de Lattre de Tassigny,
75775 PARIS CEDEX 16, France
diday@ceremade.dauphine.fr

**Mohamed Djedour**
M.S.T.D., Université USTHB,
Alger, Algeria
lmstd@usthb.dz

**Frédéric Ferraty**
Université de Toulouse, France
ferraty@cict.fr

**Nicolas Fischer**
Laboratoire National de Métrologie et d'Essais,
LNE, 1 rue Gaston Boissier, Paris, France
nicolas.fischer@lne.fr

**Françoise Fogelman Soulié**
KXEN, 25 quai Galliéni
92158 Suresnes Cedex, France
francoise.fogelman@kxen.com

**Jean-Louis Foulley**
INRA-GABI-PSGEN,
Department of Animal Genetics,
Domaine de Vilvert,
78350 Jouy-en-Josas, France
jean-louis.foulley@jouy.inra.fr

**Alexander Gammerman**
Computer Learning Research Centre,
Royal Holloway, University of London,
England
alex@cs.rhul.ac.uk

**Adam Ganz**
Department of Media Arts,
Royal Holloway, University of London, Egham
TW20 0EX, England
adam.ganz@rhul.ac.uk

**Mireille Gettler Summa**
Université Paris Dauphine,
CEREMADE, CNRS,
Place du Mal. de Lattre de Tassigny,
75016 Paris, France
summa@ceremade.dauphine.fr

**Bernard Goldfarb**
Université Paris Dauphine,
CEREMADE, CNRS,
Place du Mal. de Lattre de Tassigny,
75016 Paris, France
goldfarb@dauphine.fr

**Wenceslao Gonzàlez Manteiga**
Universidad de Santiago
de Compostela, Spain
wenceslao.gonzalez@usc.es

**David J. Hand**
Imperial College, London, England
d.j.hand@imperial.c.uk

**Frédéric Lebaron**
CURAPP, Université de Picardie
Jules Verne - CNRS, UMR 6054, France
flebaron@yahoo.fr.

**Yves Lechevallier**
Project AxIS-INRIA, Domaine de Voluceau,
Rocquencourt, B.P. 105
78153 Le Chesnay Cedex, France
yves.lechevallier@inria.fr

**Erik Marcadé**
KXEN, 25 quai Galliéni
92158 Suresnes Cedex, France
erik.marcade@kxen.com

**Mahendra Mariadassou**
Université Paris Descartes, Paris 5, MAP5
45 rue des Saints Pères, 75270 Paris Cedex
06, France
mahendra.mariadassou@jouy.inra.fr

**Adela Martínez Calvo**
Universidad de Santiago
de Compostela, Spain
adela.martinez@usc.es

**Fionn Murtagh**
Science Foundation Ireland, Dublin, Ireland
&
Department of Computer Science,
Royal Holloway, University of London,
England
fmurtagh@acm.org

**Ilia Nouretdinov**
Computer Learning Research Centre,
Royal Holloway University of London,
England
ilia@cs.rhul.ac.uk

**Dmitry Pechyony**
Akamai Technologies, Cambridge, MA, USA
dpechyon@akamai.com

**Joe Reddington**
Department of Computer Science,
Royal Holloway, University of London, Egham
TW20 0EX, England
joe.reddington@gmail.com

**Gilbert Saporta**
Chaire de Statistique Appliquée & CEDRIC,
CNAM, 292 rue Saint Martin,
Paris, France
saporta@cnam.fr

**Julien Sauvage**
KXEN, 25 quai Galliéni
92158 Suresnes Cedex, France
julien.sauvage@kxen.com

**Redouane Seraoui**
EDF R&D-STEP, 6 Quai Wattier,
78400 Chatou, France
redouane.seraoui@edf.fr

**Myriam Touati**
CEREMADE, CNRS UMR 7534 &
Université Paris-Dauphine,
Place du Mal. de Lattre de Tassigny,
75775 PARIS CEDEX 16, France
touati@ceremade.dauphine.fr

**Vladimir Vapnik**
NEC Laboratories, Princeton, NJ, USA &
Center for Computational Learning Systems,
Columbia University, New York,
NY, USA
vlad@nec-labs.com

**Maurizio Vichi**
Università La Sapienza di Roma, Italy
maurizio.vichi@uniroma1.it

**Philippe Vieu**
Université Paul Sabatier, Toulouse, France
philippe.vieu@math.univ-toulouse.fr

# Part I

# Statistical and Machine Learning

# 1

# *Mining on Social Networks*

Benjamin Chapus, Françoise Fogelman Soulié, Erik Marcadé, and Julien Sauvage

## CONTENTS

## 1.1 Introduction

The basic assumption in data mining is that observations are i.i.d. (independent and identically distributed), which is often not the case when observations are not independent but rather are interconnected, such as in interacting individuals. Social network analysis (SNA) brings ways to handle this nonindependence by exploiting the structure of connections between observations.

Originated in graph theory (Erdős & Rényi 1959 [111]), social network research has recently been developing very fast. Initially developed for sociology (Milgram et al. 1967, 1994 [231, 311]), work then focused on the analysis of (usually) small groups of individuals interacting in specific "social" ways. Then research branched out in many different areas from physics to computer science to data mining and social network techniques, which are nowadays being used in many applications: web (Huberman et al. 1999, 2001 [170, 184, 186, 185]), fraud (Pandit et al. 2007, 2011 [132, 250]), terrorism (Ressler 2006 [269]), marketing (Leskovec 2007 [203, 219]), and others. The recent development of Web 2.0 and its boom in social networking Web sites and communities (Facebook, Twitter, etc.) have generated an increased interest in social networks and their analysis (Jansen et al. 2009, 2010 [10, 180]).

**FIGURE 1.1**

A graph (left); the circle of a node $A$ (second panel); a path between nodes $A$ and $B$ (third panel) of length 4 (dotted line) and 5 (thick line); triangles (right).

There are many reasons for this growing interest, apart from the attraction of social networking sites, but in this chapter we will concentrate on what SNA brings for improving data mining modeling on *networked data* – or data that are related in a social graph – as well as for visualizing and exploring this data.

We first define what we mean by social network (Section 1.2); we introduce our approach for modeling networked data in Section 1.3; we then describe a few applications (Section 1.4); and we conclude by presenting open problems and further research (Section 1.5).

## 1.2  What Is a Social Network?

Social Network theory takes its roots in two disciplines: graph theory, first developed in the 1950s by Paul Erdős (Erdős & Rényi 1959 [111]); and sociology, which has actively investigated interaction structures in the last 50 years (Erdős & Rényi 1959 [111]). More recently, scientists from physics, computer science, and data mining ([13, 74, 170, 242, 313]), for example, have started investigating social networks, studying their structure, their dynamics, their way to propagate information, and other properties.

### 1.2.1  Definitions

A *social network* is defined as a graph, i.e., a pair $G = (V, E)$, where $V$ is a set of entities (the *nodes* or *vertices*) and $E$ the set of *edges* (the *links* or *connections*) between those entities. A graph is usually represented by a graphical representation such as, for example, Figure 1.1, which shows a graph with ten nodes (in black) connected by links. Some graphical representations might be better adapted than others for visualizing a given phenomenon, and graph visualization is a very active domain, not discussed here.

Entities can be of any sort: customers, individuals, web pages, bank accounts, credit cards, merchants, products, and so on, and links represent the interactions between these entities, with the adequate semantics for this interaction. The graph might thus represent friendship in Facebook, phone calls in a telco company, hyperlinks on the Web, and credit cards purchasing at merchants.

Graphs can be *directed* or not (making a distinction between a link from $A$ to $B$ or not); *weighted* (the links are labeled: for example, by the number of messages $A$ sent to $B$).

The motivation for studying social networks in data mining comes from the following observation: data mining relies heavily on the *i.i.d. assumption* (independent and identically distributed). To build a model, one assumes that observations are drawn independently from the same fixed (unknown) distribution. In many applications, these observations are, for example, customers who are certainly not independent: they talk to each other and thus they influence each other. We can then describe these interactions through a social network structure and the question then arises of whether and how we can build data mining models taking this structure into account. We will present our approach in Section 1.4. First, we will need to define some more concepts from social networks.

The *neighborhood* – or *circle* – of a node is the set of nodes to which that node is connected (its *neighbors*, or its *friends*). The *degree* of a node is the number of its neighbors (see Figure 1.1: node $A$ has degree 3).

A *path* is a sequence of nodes and edges linking two nodes: the length of a path is the number of arcs traversed. There can be various paths, of different lengths, between two nodes (Figure 1.1).

The *distance* (or *degree of separation*) between two nodes is the length of the shortest path.

The *diameter* of a graph is the largest distance between any two nodes in the graph. A small diameter means that the graph is compact and time needed to travel from one node to any other node is short.

A *triangle* is a set of three nodes of $G$, fully connected (containing all possible edges). The number of triangles of a node $A$ is the number of distinct triangles this node is part of: for example, in Figure 1.1, node $A$ has six neighbors, five of the fifteen possible pairs of these neighbors are connected (dotted lines); node $A$ thus has 5 triangles.

The *density* of a graph is the ratio of the number of its links to the total number possible.

The *adjacency matrix* of a graph is the matrix $(A_{ij})$ where $A_{ij} = 1$ if nodes $i$ and $j$ are connected (0 otherwise). The *weighted adjacency matrix* has $A_{ij} = w_{ij}$, where $w_{ij}$ is the weight of the edge between $i$ and $j$.

A social network is a graph, but not all graphs are social networks. Social networks have specific properties:

1. Social networks are usually big: millions of nodes, and millions, even billions of edges;

2. Nodes are usually *multilabeled*: a node can have various attributes (name, address, possession of a product, etc.);

3. Links can be of various sorts (friend, family) and carry one or more labels (in a weighted graph, there is only one label, the *weight*, indicating the strength of the connection: the number of times these two friends talked during that period for example);

4. *Homophily*: similar nodes (with similar attributes) tend to be connected;

5. *Communities*: the network can be decomposed into loosely highly interconnected groups;

6. Structural properties are specific: including the small world and scale-free properties (Watts 2003 [313]).

   - A graph is said to have the *small world property* if the average shortest path length between any two nodes is small. Stanley Milgram (Milgram 1967 [231]) developed an experiment (passing a message along a chain) in his thesis in 1967, demonstrating a small average distance of 6. He showed that *if a person is one step away from each person they know and two steps away from each person who is known by one of the people they know, then everyone is at most six steps away from any other person on Earth*. This is also known as the *six degrees of separation* property.

   - A graph is said to be *scale-free* if the distribution of its degrees follows a Pareto law (Barabasi 2002 [13]).

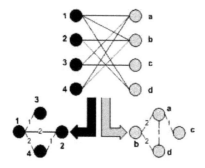

**FIGURE 1.2**

A bipartite graph connecting nodes of two disjoint sets (black and gray). The projected networks (bottom) of black nodes and gray nodes.

$$P(\text{degree} = k) \sim k^{-\gamma} \qquad (1.1)$$

There are various types of graphs:

- A *complete graph* contains all possible edges: each pair of vertices is joined by an edge;

- A *bipartite graph* (or bigraph) is a graph whose vertices can be divided into two disjoint sets U and V and edges can only connect a vertex in one set to a vertex in the other set. *Derived graphs* can then be created out of bipartite graphs: two nodes of $U$ (respectively $V$) are linked if they are both connected to at least $N_V$ nodes in $V$ (respectively $N_U$ nodes in $U$). A weight in the connections of these derived graphs could indicate the number of common nodes (see Figure 1.2).

### 1.2.2   Metrics

Many metrics have been defined to measure social network characteristics (see for example Wasserman et al. 1994, Watts 2003 [311, 313] for more details).

**Centrality metrics**

*Centrality* aims at giving an indication of the social power of a node based on how well it "connects" the network. The two most common centrality measures are:

- *Degree Centrality*: this is the number of links incident upon a node (i.e., the number of neighbors a node has, i.e., its degree);

- *Betweenness*: this is the extent to which a node lies *between* other nodes in the network. It reflects the number of people whom a person is connecting indirectly through their direct links. For graph $G = (V, E)$ with $n$ vertices in $V$, the betweenness $C_B(v)$ of vertex $v$ is:

$$C_B(v) = \sum_{s \neq v \neq t \in V, s \neq t} \frac{\sigma_{st}(v)}{\sigma_{st}} \qquad (1.2)$$

where $\sigma_{st}$ is the number of shortest paths from vertex $s$ to vertex $t$, and $\sigma_{st}(v)$ is the number of shortest paths from $s$ to $t$ that pass through vertex $v$. This measure takes into account the connectivity of the node's neighbors, giving a higher value for nodes that bridge clusters. A node with high betweenness has great influence over what flows – and does not – in the network.

## Clustering metrics

- *Local clustering coefficient*: this quantifies how close the neighbors of a given node are to being a complete graph, i.e., how well interconnected they are. The local clustering coefficient of a node $A$ is the number of triangles (*Triangle Count*) linked to $A$ divided by the number of possible triangles linked to $A$. In Figure 1.1, node $A$ has six neighbors, five of the fifteen possible pairs of these neighbors are connected (dotted lines); node $A$ thus has 5 triangles and its clustering coefficient is $1/3$ (i.e., $5/15$).

When computing metrics on graphs, one has to be careful with the *complexity* of the computation: for example computing betweenness of all vertices in a graph involves calculating the shortest paths between all pairs of vertices on a graph. On a network of $n$ vertices and $m$ edges, the runtime is $O(n^3)$, which will not permit to handling graphs with a few million vertices, in reasonable time.

## 1.2.3 Communities

No general theoretical definition of communities exists, but here, we will define a *community* as a densely connected subset of nodes that is only sparsely linked to the rest of the network.

Community analysis of a social network can be seen as a clustering of the network, based on some network-driven metric to pair nodes. But, whereas clustering is based on a distance or similarity of nodes (using their attributes), community analysis is based upon the graph structure, without taking into account the nodes attributes. Another difference is that social real-world networks being very large (500 million nodes for Facebook), the *scalability* of the algorithm becomes a major challenge.

Finding communities within an arbitrary network can be *very hard*. The number of communities in the network is typically unknown and the communities are often of unequal size and/or density and/or cohesion. Detected communities shall be as close as possible to cliques (complete subgraphs). Because of these difficulties, many methods for community identification have been developed and employed with varying success. Traditional methods used in graph theory, such as hierarchical clustering, usually perform poorly for large networks. Hence, most community detection methods used for complex networks have been developed during the last decade using specific metrics (see Fortunato 2010 [133] for a very detailed survey):

- *Girvan-Newman algorithm* (Girvan & Newman 2002 [143]): identifies edges (with the betweenness measure), which lie between communities and then removes them, leaving behind just the communities themselves. The algorithm returns results of reasonable quality but runs very slowly, taking time $O(m^2 n)$ on a network of $n$ vertices and $m$ edges, making it impractical for networks of more than a few thousand nodes.

- *Hierarchical clustering* (Clauset et al. 2004 [74]): a similarity measure between nodes is defined (for example the Jaccard index, or the Hamming distance between rows of the adjacency matrix of the graph). Similar nodes are grouped into communities according to this measure.

- *Clauset-Newman-Moore (CNM) algorithm* (Clauset et al. 2008 [73]): it is based on a measure of network clustering called *modularity*. The algorithm is a bottom-up agglomerative clustering that continuously finds and merges pairs of clusters trying to maximize modularity of the community structure in a greedy manner. Unfortunately, CNM algorithm does not scale well and its use is practically limited to networks whose sizes are up to 500,000 nodes.

  – Modularity was introduced by Girvan and Newman in 2004 (Girvan & Newman 2002 [143]): it compares the number of links inside communities to the expected number of links in a random network. It computes the numbers of links outside communities, and inside communities, taking into account the expected number of such edges, and can be expressed as:

$$Q = \frac{1}{2m} \sum_{i,j} \left[ A_{ij} - \frac{k_i k_j}{2m} \right] \delta(c_i, c_j) \tag{1.3}$$

  where

  $A_{ij}$ is the weighted adjacency matrix;
  $k = \sum_j A_{ij}$ is the sum of weights of arcs attached to $i$;
  $c_i$ is the community of node $i$;
  $\delta(u, v) = 1$ if $u = v$, and 0 otherwise;
  $m = \frac{1}{2} \sum_{i,j} A_{ij}$.

  – Note that $\frac{k_i k_j}{2m}$ is the expected number of edges between the nodes with degree $k_i$ and $k_j$ in an $m$ edges network.

- *Modularity optimization methods*: these methods detect communities by searching over possible divisions of a network for those having a particularly high modularity. Since an exhaustive search over all possible divisions is usually intractable, practical algorithms are based on approximate optimization methods such as greedy algorithms (CNM), or simulated annealing, with different approaches offering different balances between speed and accuracy, scalability being the biggest issue.

  – Wakita algorithm (Wakita & Tsurumi 2007 [309]): the authors propose a variation of the CNM algorithm, based on three kinds of metrics (consolidation ratio) to control the process of community analysis trying to balance the sizes of the communities. Wakita algorithm can be used for networks of moderate size (a few million nodes at most).

  – Louvain algorithm (Blondel et al. 2008 [27]): it is faster and more scalable than Wakita. The algorithm is divided in two phases that are repeated iteratively:

---

**PHASE 1**

1. Assign a different community to each node in the network;
2. While Modularity Gain $> \epsilon$:
   For each node $i$ in network
   * Remove $i$ from its community;
   * For each neighbor $j$ of $i$:
     Assign $i$ to community of $j$; Compute Modularity Gain and keep the best assignment.

This first phase stops when a local maximum of the modularity is attained, i.e., when no individual move can improve the modularity.

**PHASE 2**

This consists in building a new network whose nodes are the communities found during the first phase. Note that the weights of the links between the new nodes are given by the sum of the weights of the links between nodes in the corresponding two communities. Go back to PHASE 1.

The passes are iterated until there is no more change and a (local) maximum of modularity is attained.

---

Part of the algorithm efficiency results from the fact that the gain in modularity $\Delta Q$ obtained by moving an isolated node $i$ into a community $C$ can be very easily locally computed (see (Blondel et al. 2008 [27])). The algorithm seems to outperform all other known community detection methods in terms of computation time, with a good quality of detection in terms of modularity and a final hierarchical community structure (phase 2 can be iterated).

It was recently shown that modularity optimization suffers from limitations: in (Fortunato & Barthelemy 2007 [134]), it was demonstrated that modularity-based community detection algorithms have a *resolution limit*, such that small communities in a large network are invisible. In (Kumpula 2008 [192]) other limitations are pointed out.

– *Clique percolation* (Palla et al. 2005 [249]): the k-clique percolation method represents a completely different approach and has some very interesting potential. Intuitively, it is reasonable to argue that most triangles (or k-cliques, to some extent) are confined *inside communities*. In addition, the larger k is, the more strictly the k-cliques are restricted to communities because connections between communities are sparse. The clique percolation method relies on these observations and on the notion that members of a community should be reached through densely intraconnected subsets of nodes. Compared to most alternatives, the main advantage of this method is that it allows *overlaps* between the communities. But on the other hand it is not designed for hierarchical community structure.

– *Other algorithms*: there are many other algorithms ([133, 259]), some especially designed for taking into consideration the dynamic evolution of Communities (Lin et al. 2008 [209]).

We will now turn to the implementation of social networks in KXEN, Inc. and see how it can be used for data mining.

---

## 1.3   KXEN's Approach for Modeling Networked Data

KXEN has developed a Social Network Analysis module (KSN, KXEN Social Network), with the following objectives:

- Allow to extract as many social networks as necessary from transaction data (in only one pass through the dataset);

**TABLE 1.1**

Contact data used to build social networks.

| Node 1 | Node 2 | Date | Type | Duration |
|---|---|---|---|---|
| 33 1 35 24 42 18 | 33 6 75 36 21 13 | 07-Jan-10 | Voice | 00:04:22 |
| 33 6 75 36 21 13 | 33 6 21 43 27 11 | 10-Jan-10 | SMS | 00:00:10 |
| ... | | | | |
| ... | | | | |

- Create many metrics on the social network and extract communities, in – if possible – almost linear time;

- Allow to easily integrate social attributes (metrics and communities indices) into customers' database, so that they can be further used to build predictive models.

The method that KXEN has implemented is described in Figure 1.3 below (see also Fogelman Soulié 2010 [130] for a video presentation of these ideas). We start from *contact data*, which describe entities and connections. In Table 1.1 above, we show, for example, data from a telco operator: entities are customers identified by their phone number. Each time number $i$ calls number $j$, a row is created in a table (the *Call Detail Record* table) indicating the two numbers, the date of the call, its duration, the type of call, and so on. In banking, the table could contain bank accounts and money transfers; credit card purchasing at merchants (this would give a bipartite graph); in retail, the table could hold loyalty cardholders purchasing products; and in social networking sites, it would be internauts "liking" friends. Obviously, many situations in industry can be described by such contact data: defining what is meant by "contact," collecting the data, aggregating it, and so on will represent a major effort and must be tuned to the particular business objective.

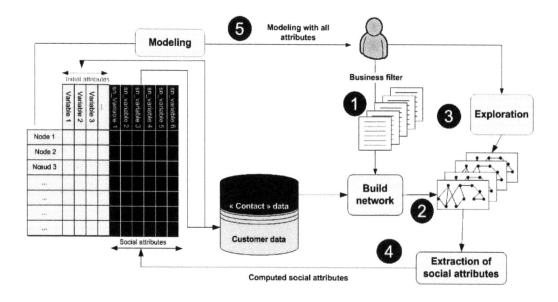

**FIGURE 1.3**

KXEN method for building and mining social networks.

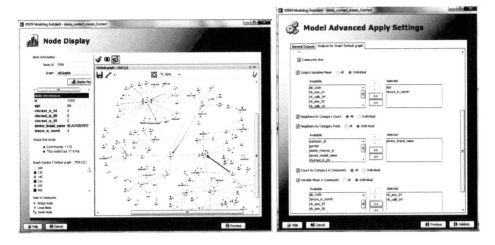

**FIGURE 1.4**
Exploring a social network with KSN (left); generating social variables with KSN (right).

1. We then define various *business filters* to specify the semantics of the interaction. For example, in the telco example above, we could have a filter setting a connection only on voice calls of at least a certain duration (1 minute) during working days and working hours (this would presumably select business calls). Another filter could select only voice calls made during weekends or working days between 6 p.m. and 8 a.m. (selecting friends and family calls). With only one contact dataset, we can thus define many different business filters, depending on the problem we want to solve. However, the results obtained will indeed heavily depend upon the social network built and different social networks, built from the very same original contact dataset might help answer very different questions [217].

2. For each of these business filters, build the associated social network. KSN can build all the social networks for the different business filters through reading the contact dataset only once (which is critical since the contact dataset might include billions of lines). Various variables characterizing the network are computed:

   - Density, degree distribution and power law exponent (equation 1.1), degree median;
   - Communities: number of detected communities, modularity, intra/interlinks median.

3. Explore the networks: KSN provides a graphical interface to explore the network "bottom-up." Starting from a given node, one can progressively visualize its neighbors, its neighbors' neighbors, their community index, and so on (Figure 1.4 left).

4. Extract *social variables*: we now use the network to compute new variables, attached to each node, including (Figure 1.4 right):

   - *Structure social variables*
     - Degree, number of triangles;
     - On the node community (if any): node community index.

- *Local social attributes*: if the nodes are labeled by some attributes (for example, customer data), then, for each node, we compute social attributes based upon the attributes of the node's neighbors (circle) or the node's community members (community).
  - For each continuous attribute, the mean of the neighbors attribute;
  - For each categorical attribute, for each of its values, the number of neighbors that have that value; the ratio of neighbors that have that value to the total number of neighbors.

  The number of social variables, which can be generated from one social network, can thus be very large (a few hundred), depending upon the number of attributes used for each node.

5. Build a model using the original entities attributes enriched with the social variables. In KXEN, models are built using a regularized regression [131], implemented through Vapnik's Structural Risk Minimization approach [300, 301]. This implementation allows building models using arbitrarily large numbers of variables (thousands), even if they are correlated. When we add social variables, we will typically add many (correlated) variables: however, these variables usually bring a different sort of information than the original attributes, resulting in models with better performances, as will be shown in Section 1.4.

The approach we have described is one way to exploit the interdependency between the observations. It is of course not the optimal solution and many authors have proposed other approaches. However, we still build our models assuming that the i.i.d. assumption holds for our entities, but by capturing some of the dependency into the social variables, we can hope to improve our models. We will now demonstrate this in a few examples in the next section.

## 1.4    Applications

Social network analysis is now being used in various industrial applications: in marketing, recommendations, brand management, fraud detection and antimoney laundering, fight against terrorism, and so on.

We describe here a few examples, illustrating some of the ways social network techniques can improve data mining models.

### 1.4.1    Telecommunications

The telecommunications domain is certainly the industrial domain that has most used Social Network Analysis up to now. This is due to the fact that information already available is well-structured in a form immediate for Link analysis (see Table 1.1), and that a graphical display based on links is an intuitive representation for a typical telco phenomenon (*calls*), whether these links are directed or not, and weighted or not. Call Detail Records are stored in databases, usually saved for 1 to 3 months, and usually consist of terabytes of data.

From these CDRs, one can define *various semantics of interaction*, and thus build many different social networks, depending on the product [voice, Short Message Service (SMS), Multimedia Messaging Service (MMS)], the intensity (number of communications, the period, geography, etc.

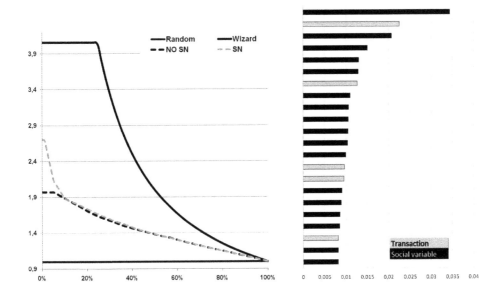

**FIGURE 1.5**
The figure shows the improvement in lift (left) and contribution of variables (right) when using social variables in a telecommunications application of churn detection.

Applying the modeling procedure described in Section 1.3, we can create extrapredictive variables derived from the social network structure and the attributes of *On-net* nodes, i.e., customers of the telco operators, and add them to the explanatory variables. We can then build churn models (i.e., predicting which customers will switch to another telco operator), propensity models, and so on, and should expect an increased lift of the models, hence improving global marketing performance.

In the example of Figure 1.5, coming from a large telco operator in Europe, we can see that including social variables increases the lift of a churn model in the first decile (+47% in first percentile) and that social variables (in black in the figure) are among the most contributive variables (actually out of the total 460 variables, the 214 social variables bring 46% of the total contribution; of the top 20 most contributive variables, 15 are social variables, bringing over 76% of the total contribution).

## 1.4.2  Fraud

Credit card transactions on the Internet have grown steadily these last ten years, spurring an accompanying strong increase in fraud. We have shown [132], in the context of a research project funded by French science foundation ANR (Agence Nationale de la Recherche project eFraudBox) that using social network techniques in this context can significantly improve fraud detection and analysis.

From the transaction data of purchases by credit cardholders at merchants, one can build a bipartite network (as in Figure 1.2) of credit cards and merchants. When deriving the simple networks of merchants (for example), one can then compute social variables attached to the merchant (as described in Section 1.3 above). Each transaction is then enriched with these new social variables.

As can be seen in Figure 1.6, coming from the work in cFraudBox (with data from GIE

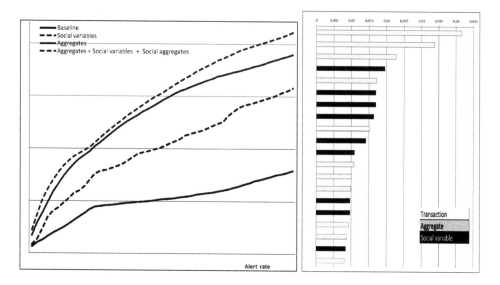

**FIGURE 1.6**
The figure shows the improvement in sensitivity (left) and contribution of variables (right) when using transaction data, aggregates, and social variables (in black) in a model of fraud detection in Internet credit card purchase.

Cartes Bancaires), we can see that including social variables increases the sensitivity of a detection model for a very small alert rate (the exact figures are not shown for confidentiality reasons) and that social variables (in black in the figure) are among the most contributive variables (actually of the top 20 most contributive variables, 9 are social variables, bringing over 40% of the total contribution).

### 1.4.3   Retail

In retail, one can use transaction data from purchases: a user buys a product at a certain date, in some quantity. From this data, we build a bipartite network (as in Figure 1.2) of customers and products. One can then derive the customers' network and the products' network. If one then computes the communities in each of these networks, one will obtain a *co-clustering* of products and customers: groups of customers buying preferably certain groups of products.

In a research project funded by ANR, CADI, we have worked on purchase data at La Boîte à Outils (a large chain of home improvement stores in France). We selected customers spending a relatively large amount of money in short periods (these customers are running a *project*: for example, redoing their kitchen). When computing the communities in the customers and products networks, we find communities of products such as those shown in Figure 1.7, showing grouping of products relevant for a given project (installing a pool, building a fence, and so on). Now communities of customers and products are like shown in the figure: one customers' community will buy most of its products in a products community.

We then built a recommendation model using Collaborative Filtering on each of these customers' community, thus in effect building a model specialized for the type of project those customers have in mind: for each customer, depending on his past purchases, we provide recommendations of the products he is most likely to be interested in. As Figure 1.7 (right) shows, all performance measures are significantly increased when using the

| Community | Product |
|---|---|
| 10 | Pool cleaning |
| 10 | Pool accessories |

| Community | Product |
|---|---|
| 15 | Wires in rolls |
| 15 | Poles & Accessories |

| | CP1 | CP2 | CP3 |
|---|---|---|---|
| CC1 | 7 | 60 | 3 |
| CC2 | 0 | 3 | 46 |

| | Precision | Recall | F1 |
|---|---|---|---|
| Transactions | 10.85 % | 9.57 % | 8.77 % |
| Communities | 11.14 % | 9.87 % | 9.03 % |

**FIGURE 1.7**

The figure shows the communities found in the products network (left); the co-clustering (middle) and the improvement of using a model personalized on communities in a recommendation application in retail (right; all error bars are about 0.1%).

communities-based model as compared to the simple transaction-based Collaborative Filtering model.

## 1.5   Conclusion

We have presented a methodology by which one can exploit the interaction structure of observations to take into account this structure and improve data mining models. This is certainly only the very beginning of this line of research and lots of open problems remain: how to include the dynamics of social graphs, how to predict the diffusion of information in such networks, and so on.

More developments will be needed if one wants to build models on data that violate the i.i.d. assumption.

## Acknowledgments

The authors thank ANR (Agence Nationale de la Recherche) for funding the eFraudBox project and the CADI project referred to in this chapter. GIE Cartes Bancaires, partner in eFraudBox, kindly provided the data used in the fraud section (Section 1.4.2). Savaneary Sean (KXEN) provided the results in Section 1.4.3.

## Glossary

**Co-clustering:** building clusters of two sets of elements at the same time (for example, the two sets of nodes in a bipartite network, by extracting communities in the derived networks).

**Communities:** subsets of highly interconnected nodes in a social network, these subsets being loosely interconnected.

**i.i.d. (independent and identically distributed):** basic assumption of data mining, which states that observations form a sample of independent observations drawn from the same underlying distribution.

**Social network:** a structure comprising entities (nodes) interconnected by links. In effect, social network is a graph, but with specific characteristics.

**Social network analysis:** a set of techniques to describe and analyze the structure, the dynamics, and the behavior of social networks.

# 2

## Large-Scale Machine Learning with Stochastic Gradient Descent

Léon Bottou

## CONTENTS

## 2.1 Introduction

The computational complexity of learning an algorithm becomes the critical limiting factor when one envisions very large datasets. This contribution advocates stochastic gradient algorithms for large-scale machine learning problems. The first section describes the stochastic gradient algorithm. The second section presents an analysis that explains why stochastic gradient algorithms are attractive when the data is abundant. The third section discusses the asymptotical efficiency of estimates obtained after a single pass over the training set. The last section presents empirical evidence.

## 2.2 Learning with Gradient Descent

Let us first consider a simple supervised learning setup. Each example $z$ is a pair $(x, y)$ composed of an arbitrary input $x$ and a scalar output $y$. We consider a *loss function* $\ell(\hat{y}, y)$ that measures the cost of predicting $\hat{y}$ when the actual answer is $y$, and we choose a family $\mathcal{F}$ of functions $f_w(x)$ parametrized by a weight vector $w$. We seek the function $f \in \mathcal{F}$ that minimizes the loss $Q(z, w) = \ell(f_w(x), y)$ averaged on the examples. Although we would like to average over the unknown distribution $dP(z)$ that embodies the Laws of Nature, we must often settle for computing the average on a sample $z_1 \ldots z_n$.

$$E(f) = \int \ell(f(x), y) \, dP(z) \qquad E_n(f) = \frac{1}{n} \sum_{i=1}^{n} \ell(f(x_i), y_i) \qquad (2.1)$$

The *empirical risk* $E_n(f)$ measures the training set performance. The *expected risk* $E(f)$ measures the generalization performance, that is, the expected performance on future examples. The statistical learning theory (Vapnik & Chervonenkis 1971 [302]) justifies minimizing the empirical risk instead of the expected risk when the chosen family $\mathcal{F}$ is sufficiently restrictive.

### 2.2.1   Gradient Descent

It has often been proposed (e.g., Rumelhart et al. 1986 [278]) to minimize the empirical risk $E_n(f_w)$ using *gradient descent* (GD). Each iteration updates the weights $w$ on the basis of the gradient of $E_n(f_w)$,

$$w_{t+1} \;=\; w_t - \gamma \frac{1}{n} \sum_{i=1}^{n} \nabla_w Q(z_i, w_t), \qquad (2.2)$$

where $\gamma$ is an adequately chosen gain. Under sufficient regularity assumptions, when the initial estimate $w_0$ is close enough to the optimum, and when the gain $\gamma$ is sufficiently small, this algorithm achieves *linear convergence* (Dennis & Schnabel 1983 [88]), that is, $-\log \rho \sim t$, where $\rho$ represents the residual error.

Much better optimization algorithms can be designed by replacing the scalar gain $\gamma$ by a positive definite matrix $\Gamma_t$ that approaches the inverse of the Hessian of the cost at the optimum:

$$w_{t+1} \;=\; w_t - \Gamma_t \frac{1}{n} \sum_{i=1}^{n} \nabla_w Q(z_i, w_t). \qquad (2.3)$$

This *second-order gradient descent* (2GD) is a variant of the well-known Newton algorithm. Under sufficiently optimistic regularity assumptions, and provided that $w_0$ is sufficiently close to the optimum, second-order gradient descent achieves *quadratic convergence*. When the cost is quadratic and the scaling matrix $\Gamma$ is exact, the algorithm reaches the optimum after a single iteration. Otherwise, assuming sufficient smoothness, we have $-\log \log \rho \sim t$.

### 2.2.2   Stochastic Gradient Descent

The *stochastic gradient descent* (SGD) algorithm is a drastic simplification. Instead of computing the gradient of $E_n(f_w)$ exactly, each iteration estimates this gradient on the basis of a single randomly picked example $z_t$:

$$w_{t+1} \;=\; w_t - \gamma_t \nabla_w Q(z_t, w_t). \qquad (2.4)$$

The stochastic process $\{ w_t, \ t=1,\ldots \}$ depends on the examples randomly picked at each iteration. It is hoped that (2.4) behaves like its expectation (2.2) despite the noise introduced by this simplified procedure.

Since the stochastic algorithm does not need to remember which examples were visited during the previous iterations, it can process examples on the fly in a deployed system. In such a situation, the stochastic gradient descent directly optimizes the expected risk, since the examples are randomly drawn from the ground truth distribution.

**TABLE 2.1**
Stochastic gradient algorithms for various learning systems.

| Loss | Stochastic algorithm |
|---|---|
| **Adaline** (Widrow & Hoff 1960 [315]) $Q_{\text{adaline}} = \frac{1}{2}\left(y - w^\top \Phi(x)\right)^2$ $\Phi(x) \in \mathbb{R}^d,\ y = \pm 1$ | $w \leftarrow w + \gamma_t \left(y_t - w^\top \Phi(x_t)\right)\Phi(x_t)$ |
| **Perceptron** (Rosenblatt 1957 [273]) $Q_{\text{perceptron}} = \max\{0, -y\, w^\top \Phi(x)\}$ $\Phi(x) \in \mathbb{R}^d,\ y = \pm 1$ | $w \leftarrow w + \gamma_t \begin{cases} y_t\, \Phi(x_t) & \text{if } y_t\, w^\top \Phi(x_t) \leq 0 \\ 0 & \text{otherwise} \end{cases}$ |
| **K-Means** (MacQueen 1967 [218]) $Q_{\text{kmeans}} = \min_k \frac{1}{2}(z - w_k)^2$ $z \in \mathbb{R}^d,\ w_1 \dots w_k \in \mathbb{R}^d$ $n_1 \dots n_k \in \mathbb{N},\ \text{initially } 0$ | $k^* = \arg\min_k (z_t - w_k)^2$ $n_{k^*} \leftarrow n_{k^*} + 1$ $w_{k^*} \leftarrow w_{k^*} + \frac{1}{n_{k^*}}(z_t - w_{k^*})$ |
| **SVM** (Cortes & Vapnik 1995 [75]) $Q_{\text{svm}} = \lambda w^2 + \max\{0, 1 - y\, w^\top \Phi(x)\}$ $\Phi(x) \in \mathbb{R}^d,\ y = \pm 1,\ \lambda > 0$ | $w \leftarrow w - \gamma_t \begin{cases} \lambda w & \text{if } y_t\, w^\top \Phi(x_t) > 1, \\ \lambda w - y_t\, \Phi(x_t) & \text{otherwise.} \end{cases}$ |
| **Lasso** (Tibshirani 1996 [292]) $Q_{\text{lasso}} = \lambda |w|_1 + \frac{1}{2}\left(y - w^\top \Phi(x)\right)^2$ $w = (u_1 - v_1, \dots, u_d - v_d)$ $\Phi(x) \in \mathbb{R}^d,\ y \in \mathbb{R},\ \lambda > 0$ | $u_i \leftarrow \left[u_i - \gamma_t(\lambda - (y_t - w^\top \Phi(x_t))\Phi_i(x_t))\right]_+$ $v_i \leftarrow \left[v_i - \gamma_t(\lambda + (y_t - w_t^\top \Phi(x_t))\Phi_i(x_t))\right]_+$ with notation $[x]_+ = \max\{0, x\}$. |

The convergence of stochastic gradient descent has been studied extensively in the stochastic approximation literature. Convergence results usually require decreasing gains satisfying the conditions $\sum_t \gamma_t^2 < \infty$ and $\sum_t \gamma_t = \infty$. The Robbins-Siegmund theorem (Robbins & Siegmund 1971 [271]) provides the means to establish almost sure convergence under mild conditions (Bottou 1998 [35]), including cases where the loss function is not everywhere differentiable.

The convergence speed of stochastic gradient descent is in fact limited by the noisy approximation of the true gradient. When the gains decrease too slowly, the variance of the parameter estimate $w_t$ decreases equally slowly. When the gains decrease too quickly, the expectation of the parameter estimate $w_t$ takes a very long time to approach the optimum. Under sufficient regularity conditions (e.g., Murata 1998 [233]), the best convergence speed is achieved using gains $\gamma_t \sim t^{-1}$. The expectation of the residual error then decreases with similar speed, that is, $\mathbb{E}\,\rho \sim t^{-1}$.

The *second-order stochastic gradient descent* (2SGD) multiplies the gradients by a positive definite matrix $\Gamma_t$ approaching the inverse of the Hessian:

$$w_{t+1} = w_t - \gamma_t \Gamma_t \nabla_w Q(z_t, w_t). \tag{2.5}$$

Unfortunately, this modification does not reduce the stochastic noise and therefore does not significantly improve the variance of $w_t$. Although constants are improved, the expectation of the residual error still decreases like $t^{-1}$, that is, $\mathbb{E}\,\rho \sim t^{-1}$, (e.g., Bordes et al. 2009 [30], appendix).

### 2.2.3 Stochastic Gradient Examples

Table 2.1 illustrates stochastic gradient descent algorithms for a number of classic machine learning schemes. The stochastic gradient descent for the Perceptron, for the Adaline, and for $k$-means match the algorithms proposed in the original papers. The support vector machine (SVM) and the Lasso were first described with traditional optimization techniques. Both

$Q_{\text{svm}}$ and $Q_{\text{lasso}}$ include a regularization term controlled by the hyperparameter $\lambda$. The K-means algorithm converges to a local minimum because $Q_{\text{kmeans}}$ is nonconvex. On the other hand, the proposed update rule uses second-order gains that ensure a fast convergence. The proposed Lasso algorithm represents each weight as the difference of two positive variables. Applying the stochastic gradient rule to these variables and enforcing their positivity leads to sparser solutions.

## 2.3  Learning with Large Training Sets

Let $f^* = \arg\min_f E(f)$ be the best possible prediction function. Since we seek the prediction function from a parametrized family of functions $\mathcal{F}$, let $f_{\mathcal{F}}^* = \arg\min_{f \in \mathcal{F}} E(f)$ be the best function in this family. Since we optimize the empirical risk instead of the expected risk, let $f_n = \arg\min_{f \in \mathcal{F}} E_n(f)$ be the empirical optimum. Since this optimization can be costly, let us stop the algorithm when it reaches a solution $\tilde{f}_n$ that minimizes the objective function with a predefined accuracy $E_n(\tilde{f}_n) < E_n(f_n) + \rho$.

### 2.3.1  The Tradeoffs of Large-Scale Learning

The excess error $\mathcal{E} = \mathbb{E}\big[E(\tilde{f}_n) - E(f^*)\big]$ can be decomposed in three terms (Bottou & Bousquet 2008 [36]):

$$\mathcal{E} = \mathbb{E}\big[E(f_{\mathcal{F}}^*) - E(f^*)\big] + \mathbb{E}\big[E(f_n) - E(f_{\mathcal{F}}^*)\big] + \mathbb{E}\big[E(\tilde{f}_n) - E(f_n)\big]. \tag{2.6}$$

- The approximation error $\mathcal{E}_{\text{app}} = \mathbb{E}\big[E(f_{\mathcal{F}}^*) - E(f^*)\big]$ measures how closely functions in $\mathcal{F}$ can approximate the optimal solution $f^*$. The approximation error can be reduced by choosing a larger family of functions.

- The estimation error $\mathcal{E}_{\text{est}} = \mathbb{E}\big[E(f_n) - E(f_{\mathcal{F}}^*)\big]$ measures the effect of minimizing the empirical risk $E_n(f)$ instead of the expected risk $E(f)$. The estimation error can be reduced by choosing a smaller family of functions or by increasing the size of the training set.

- The optimization error $\mathcal{E}_{\text{opt}} = E(\tilde{f}_n) - E(f_n)$ measures the impact of the approximate optimization on the expected risk. The optimization error can be reduced by running the optimizer longer. The additional computing time depends of course on the family of function and on the size of the training set.

Given constraints on the maximal computation time $T_{\max}$ and the maximal training set size $n_{\max}$, this decomposition outlines a tradeoff involving the size of the family of functions $\mathcal{F}$, the optimization accuracy $\rho$, and the number of examples $n$ effectively processed by the optimization algorithm:

$$\min_{\mathcal{F}, \rho, n} \; \mathcal{E} = \mathcal{E}_{\text{app}} + \mathcal{E}_{\text{est}} + \mathcal{E}_{\text{opt}} \quad \text{subject to} \left\{ \begin{array}{ccc} n & \leq & n_{\max} \\ T(\mathcal{F}, \rho, n) & \leq & T_{\max} \end{array} \right. \tag{2.7}$$

Two cases should be distinguished:

- *Small-scale learning problems* are first constrained by the maximal number of examples. Since the computing time is not an issue, we can reduce the optimization error $\mathcal{E}_{\text{opt}}$ to insignificant levels by choosing $\rho$ arbitrarily small, and we can minimize the estimation error by chosing $n = n_{\max}$. We then recover the approximation-estimation tradeoff that has been widely studied in statistics and in learning theory.

- *Large-scale learning problems* are first constrained by the maximal computing time. Approximate optimization can achieve better expected risk because more training examples can be processed during the allowed time. The specifics depend on the computational properties of the chosen optimization algorithm.

### 2.3.2 Asymptotic Analysis

Solving (2.7) in the asymptotic regime amounts to ensuring that the terms of the decomposition (2.6) decrease at similar rates. Since the asymptotic convergence rate of the excess error (2.6) is the convergence rate of its slowest term, the computational effort required to make a term decrease faster would be wasted.

For simplicity, we assume in this section that the Vapnik-Chervonenkis dimensions of the families of functions $\mathcal{F}$ are bounded by a common constant. We also assume that the optimization algorithms satisfy all the assumptions required to achieve the convergence rates discussed in Section 2.2. Similar analyses can be carried out for specific algorithms under weaker assumptions (e.g., Shalev-Shwartz & Srebro 2008 [284]).

A simple application of the uniform convergence results of (Vapnik & Chervonenkis 1971 [302]), gives then the upper bound

$$\mathcal{E} \;=\; \mathcal{E}_{\mathrm{app}} + \mathcal{E}_{\mathrm{est}} + \mathcal{E}_{\mathrm{opt}} \;=\; \mathcal{E}_{\mathrm{app}} + \mathcal{O}\!\left(\sqrt{\frac{\log n}{n}} + \rho\right).$$

Unfortunately the convergence rate of this bound is too pessimistic. Faster convergence occurs when the loss function has strong convexity properties (Lee et al. 1998 [202]) or when the data distribution satisfies certain assumptions (Tsybakov 2004 [295]). The equivalence

$$\mathcal{E} \;=\; \mathcal{E}_{\mathrm{app}} + \mathcal{E}_{\mathrm{est}} + \mathcal{E}_{\mathrm{opt}} \;\sim\; \mathcal{E}_{\mathrm{app}} + \left(\frac{\log n}{n}\right)^{\alpha} + \rho, \quad \text{for some } \alpha \in \left[\tfrac{1}{2}, 1\right], \qquad (2.8)$$

provides a more realistic view of the asymptotic behavior of the excess error (e.g., Massart & Bousquet 2000, 2002 [223, 54]). Since the three components of the excess error should decrease at the same rate, the solution of the tradeoff problem (2.7) must then obey the multiple asymptotic equivalences

$$\mathcal{E} \;\sim\; \mathcal{E}_{\mathrm{app}} \;\sim\; \mathcal{E}_{\mathrm{est}} \;\sim\; \mathcal{E}_{\mathrm{opt}} \;\sim\; \left(\frac{\log n}{n}\right)^{\alpha} \;\sim\; \rho. \qquad (2.9)$$

Table 2.2 summarizes the asymptotic behavior of the four-gradient algorithm described in Section 2.2. The first three rows list the computational cost of each iteration, the number of iterations required to reach an optimization accuracy $\rho$, and the corresponding computational cost. The last row provides a more interesting measure for large-scale machine learning purposes. Assuming we operate at the optimum of the approximation-estimation-optimization tradeoff (2.7), this line indicates the computational cost necessary to reach a predefined value of the excess error, and therefore of the expected risk. This is computed by applying the equivalences (2.9) to eliminate $n$ and $\rho$ from the third row results.

Although the stochastic gradient algorithms, SGD and 2SGD, are clearly the worst optimization algorithms (third row), they need less time than the other algorithms to reach a predefined expected risk (fourth row). Therefore, in the large-scale setup, that is, when the limiting factor is the computing time rather than the number of examples, the stochastic learning algorithms perform asymptotically better!

**TABLE 2.2**
Asymptotic equivalents for various optimization algorithms: gradient descent (GD, eq. 2.2), second-order gradient descent (2GD, eq. 2.3), stochastic gradient descent (SGD, eq. 2.4), and second-order stochastic gradient descent (2SGD, eq. 2.5). Although they are the worst optimization algorithms, SGD and 2SGD achieve the fastest convergence speed on the expected risk. They differ only by constant factors not shown in this table, such as condition numbers and weight vector dimension.

| | GD | 2GD | SGD | 2SGD |
|---|---|---|---|---|
| Time per iteration: | $n$ | $n$ | $1$ | $1$ |
| Iterations to accuracy $\rho$: | $\log \frac{1}{\rho}$ | $\log \log \frac{1}{\rho}$ | $\frac{1}{\rho}$ | $\frac{1}{\rho}$ |
| Time to accuracy $\rho$: | $n \log \frac{1}{\rho}$ | $n \log \log \frac{1}{\rho}$ | $\frac{1}{\rho}$ | $\frac{1}{\rho}$ |
| Time to excess error $\varepsilon$: | $\frac{1}{\varepsilon^{1/\alpha}} \log^2 \frac{1}{\varepsilon}$ | $\frac{1}{\varepsilon^{1/\alpha}} \log \frac{1}{\varepsilon} \log \log \frac{1}{\varepsilon}$ | $\frac{1}{\varepsilon}$ | $\frac{1}{\varepsilon}$ |

## 2.4   Efficient Learning

Let us add an additional example $z_t$ to a training set $z_1 \ldots z_{t-1}$. Since the new empirical risk $E_t(f)$ remains close to $E_{t-1}(f)$, the empirical minimum $w_{t+1}^* = \arg\min_w E_t(f_w)$ remains close to $w_t^* = \arg\min_w E_{t-1}(f_w)$. With sufficient regularity assumptions, a first-order calculation gives the result

$$w_{t+1}^* = w_t^* - t^{-1} \Psi_t \nabla_w Q(z_t, w_t^*) + \mathcal{O}(t^{-2}), \qquad (2.10)$$

where $\Psi_t$ is the inverse of the Hessian of $E_t(f_w)$ in $w_t^*$. The similarity between this expression and the second-order stochastic gradient descent rule (2.5) has deep consequences. Let $w_t$ be the sequence of weights obtained by performing a *single second-order stochastic gradient pass* on the randomly shuffled training set. With adequate regularity and convexity assumptions, we can prove (e.g., Bottou & LeCun 2005 [37])

$$\lim_{t \to \infty} t \left( E(f_{w_t}) - E(f_{\mathcal{F}}^*) \right) = \lim_{t \to \infty} t \left( E(f_{w_t^*}) - E(f_{\mathcal{F}}^*) \right) = \mathcal{I} > 0. \qquad (2.11)$$

Therefore, a single pass of second-order stochastic gradient provides a prediction function $f_{w_t}$ that approaches the optimum $f_{\mathcal{F}}^*$ as efficiently as the empirical optimum $f_{w_t^*}$. In particular, when the loss function is the log likelihood, the empirical optimum is the asymptotically efficient maximum likelihood estimate, and the second order stochastic gradient estimate is also asymptotically efficient.

Unfortunately, second-order stochastic gradient descent is computationally costly because each iteration (2.5) performs a computation that involves the large dense matrix $\Gamma_t$. Two approaches can work around this problem.

- Computationally efficient approximations of the inverse Hessian trade asymptotic optimality for computation speed. For instance, the stochastic gradient descent quasi-Newton (SGDQN) algorithm (Bordes et al. 2009 [30]) achieves interesting speeds using a diagonal approximation.

- The *averaged stochastic gradient descent* (ASGD) algorithm (Polyak & Juditsky

1992 [257] performs the normal stochastic gradient update (2.4) and recursively computes the average $\bar{w}_t = \frac{1}{t}\sum_{i=1}^{t} w_t$ :

$$ w_{t+1} = w_t - \gamma_t \nabla_w Q(z_t, w_t), \quad \bar{w}_{t+1} = \frac{t}{t+1}\bar{w}_t + \frac{1}{t+1}w_{t+1}. \tag{2.12} $$

When the gains $\gamma_t$ decrease slower than $t^{-1}$, the $\bar{w}_t$ converges with the optimal asymptotic speed (2.11). Reaching this asymptotic regime can take a very long time in practice. A smart selection of the gains $\gamma_t$ helps achieving the promised performance (Xu 2010 [317]).

## 2.5 Experiments

This section briefly reports experimental results illustrating the actual performance of stochastic gradient algorithms on a variety of linear systems. We use gains $\gamma_t = \gamma_0(1 + \lambda\gamma_0 t)^{-1}$ for SGD and, following (Xu 2010 [317]). $\gamma_t = \gamma_0(1 + \lambda\gamma_0 t)^{-0.75}$ for ASGD. The initial gains $\gamma_0$ were set manually by observing the performance of each algorithm running on a subset of the training examples.

Figure 2.1 reports results achieved using SGD for a linear SVM trained for the recognition of the CCAT category in the RCV1 dataset (Lewis et al. 2004 [205]) using both the hinge loss ( $Q_{\text{svm}}$ in Table 2.1), and the log loss ( $Q_{\text{logsvm}} = \lambda w^2 + \log(1 + \exp(-y\,w^\top \Phi(x)))$ ). The training set contains 781,265 documents represented by 47,152 relatively sparse TF/IDF (term frequency, inverse document frequence weight) features. SGD runs considerably faster than either the standard SVM solvers SVMLIGHT and SVMPERF (Joachims 2006 [182]) or the superlinear optimization algorithm TRON (Lin et al. 2007 [208]).

Figure 2.2 reports results achieved using SGD, SGDQN, and ASGD for a linear SVM trained on the ALPHA task of the 2008 Pascal Large Scale Learning Challenge (see Bordes et al. 2009 [30]) using the squared hinge loss ( $Q_{\text{sqsvm}} = \lambda w^2 + \max\{0, 1 - y\,w^\top\Phi(x)\}^2$ ). The

| Algorithm | Time | Test Error |
|---|---|---|
| *Hinge loss SVM, $\lambda = 10^{-4}$.* | | |
| SVMLIGHT | 23,642 s. | 6.02 % |
| SVMPERF | 66 s. | 6.03 % |
| SGD | **1.4** s. | 6.02 % |
| *Log loss SVM, $\lambda = 10^{-5}$.* | | |
| TRON (-e0.01) | 30 s. | 5.68 % |
| TRON (-e0.001) | 44 s. | 5.70 % |
| SGD | **2.3** s. | 5.66 % |

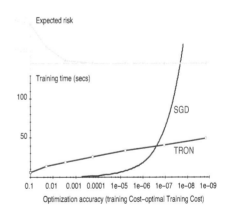

**FIGURE 2.1**
Results achieved with a linear SVM on the RCV1 task. The lower half of the plot shows the time required by SGD and TRON to reach a predefined accuracy $\rho$ on the log loss task. The upper half shows that the expected risk stops improving long before the superlinear TRON algorithm overcomes SGD.

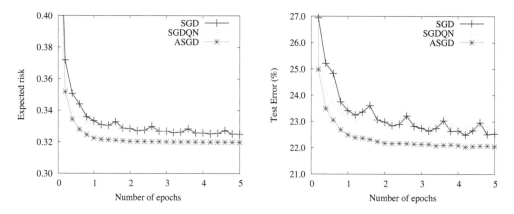

**FIGURE 2.2**

Comparison of the test set performance of SGD, SGDQN, and ASGD for a linear squared hinge SVM trained on the ALPHA task of the 2008 Pascal Large Scale Learning Challenge. ASGD nearly reaches the optimal expected risk after a single pass.

training set contains 100,000 patterns represented by 500 centered and normalized variables. Performances measured on a separate testing set are plotted against the number of passes over the training set. ASGD achieves near-optimal results after one pass.

Figure 2.3 reports results achieved using SGD, SGDQN, and ASGD for a CRF (Lafferty et al. 2001 [194]) trained on the CONLL 2000 Chunking task (Tjong et al. 2000 [293]). The training set contains 8936 sentences for a $1.68 \times 10^6$ dimensional parameter space. Performances measured on a separate testing set are plotted against the number of passes

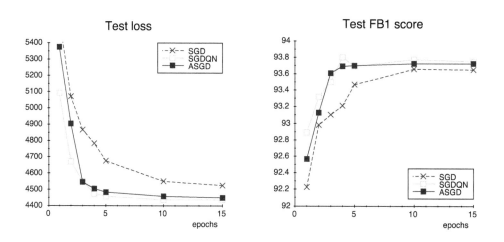

**FIGURE 2.3**

Comparison of the test set performance of SGD, SGDQN, and ASGD on a conditional random field (CRF) trained on the Conference on Computational Natual Language Learning (CONLL) Chunking task. On this task, SGDQN appears more attractive because ASGD does not reach its asymptotic performance.

over the training set. SGDQN appears more attractive because ASGD does not reach its asymptotic performance. All three algorithms reach the best test set performance in a couple minutes. The standard CRF limited-memory Broyden-Fletcher-Goldfarb-Shanno (L-BFGS) optimizer takes 72 minutes to compute an equivalent solution.

# 3

# Fast Optimization Algorithms for Solving SVM+

**Dmitry Pechyony and Vladimir Vapnik**

## CONTENTS

In the *Learning Using Privileged Information (LUPI)* model, along with standard training data in the primary space, a teacher supplies a student with additional (privileged) information in the secondary space. The goal of the learner is to find a classifier with a low generalization error in the primary space. One of the major algorithmic tools for learning in the LUPI model is support vector machine plus (SVM+). In this chapter we show two fast algorithms for solving the optimization problem of SVM+. To motivate the usage of SVM+, we show how it can be used to learn from the data that is generated by human computation games.

## 3.1 Introduction

Recently a new learning paradigm, called *Learning Using Privileged Information (LUPI)*, was introduced by Vapnik et al. (Vapnik et al. 2006, 2008, 2009 [301, 304, 305]). In this paradigm, in addition to the standard training data, $(\mathbf{x}, y) \in X \times \{\pm 1\}$, a teacher supplies a student with the *privileged information* $\mathbf{x}^* \in X^*$. The privileged information is only available for the training examples and is never available for the test examples. The privileged information paradigm requires, given a training set $\{(\mathbf{x}_i, \mathbf{x}_i^*, y_i)\}_{i=1}^n$, to find a function $f : X \to \{-1, 1\}$ with small generalization error for the unknown test data $\mathbf{x} \in X$.

The LUPI paradigm can be implemented based on the well-known SVM algorithm (Cortes & Vapnik 1995 [75]). The decision function of SVM is $h(\mathbf{z}) = \mathbf{w} \cdot \mathbf{z} + b$, where $\mathbf{z}$ is a feature map of $\mathbf{x}$, and $\mathbf{w}$ and $b$ are the solution of the following optimization problem:

$$\min_{\mathbf{w}, b, \xi_1, \ldots, \xi_n} \frac{1}{2} \|\mathbf{w}\|_2^2 + C \sum_{i=1}^{n} \xi_i \tag{3.1}$$

$$\text{s.t. } \forall \ 1 \leq i \leq n, \ y_i (\mathbf{w} \cdot \mathbf{z}_i + b) \geq 1 - \xi_i,$$

$$\forall \ 1 \leq i \leq n, \ \xi_i \geq 0.$$

Let $h' = (\mathbf{w}', b')$ be the best possible decision function (in terms of the generalization error) that SVM can find. Suppose that for each training example $\mathbf{x}_i$ an oracle gives us the value of the slack $\xi_i' = 1 - y_i(\mathbf{w}' \cdot \mathbf{x}_i + b')$. We substitute these slacks into (3.1), fix them and optimize (3.1) only over $\mathbf{w}$ and $b$. We denote such a variant of SVM as *OracleSVM*. By Proposition 1 of (Vapnik & Vashist 2009 [304]), the generalization error of the hypothesis found by OracleSVM converges to the one of $h'$ with the rate of $1/n$. This rate is much faster than the convergence rate $1/\sqrt{n}$ of SVM.

In the absence of the optimal values of slacks we use the privileged information $\{\mathbf{x}_i^*\}_{i=1}^{n}$ to estimate them. Let $\mathbf{z}_i^* \in Z^*$ be a feature map of $\mathbf{x}_i^*$. Our goal is to find a *correcting function* $\phi(\mathbf{x}_i^*) = \mathbf{w}^* \cdot \mathbf{z}_i^* + d$ that approximates $\xi_i'$. We substitute $\xi_i = \phi(\mathbf{x}_i^*)$ into (3.1) and obtain the following modification (3.2) of SVM, called SVM+ (Vapnik 2006 [301]):

$$\min_{\mathbf{w}, b, \mathbf{w}^*, d} \frac{1}{2} \|\mathbf{w}\|_2^2 + \frac{\gamma}{2} \|\mathbf{w}^*\|_2^2 + C \sum_{i=1}^{n} (\mathbf{w}^* \cdot \mathbf{z}_i^* + d) \tag{3.2}$$

$$\text{s.t.} \forall \ 1 \leq i \leq n, \ y_i (\mathbf{w} \cdot \mathbf{z}_i + b) \geq 1 - (\mathbf{w}^* \cdot \mathbf{z}_i^* + d),$$

$$\forall \ 1 \leq i \leq n, \ \mathbf{w}^* \cdot \mathbf{z}_i^* + d \geq 0,$$

The objective function of SVM+ contains two hyperparameters, $C > 0$ and $\gamma > 0$. The term $\gamma \|\mathbf{w}^*\|_2^2 / 2$ in (3.2) is intended to restrict the capacity (or VC-dimension) of the function space containing $\phi$.

Let $K_{ij} = K(\mathbf{x}_i, \mathbf{x}_j)$ and $K_{ij}^* = K^*(\mathbf{x}_i^*, \mathbf{x}_j^*)$ be kernels in the decision and the correcting space, respectively. A common approach to solve (3.1) is to consider its dual problem. We also use this approach to solve (3.2). The kernalized duals of (3.1) (3.2) are (3.3) and (3.6), respectively.

$$\max_{\boldsymbol{\alpha}} \ D(\boldsymbol{\alpha}) = \sum_{i=1}^{n} \alpha_i - \frac{1}{2} \sum_{i,j=1}^{n} \alpha_i \alpha_j y_i y_j K_{ij} \tag{3.3}$$

$$\text{s.t.} \sum_{i=1}^{n} y_i \alpha_i = 0, \tag{3.4}$$

$$\forall \ 1 \leq i \leq n, \ 0 \leq \alpha_i \leq C. \tag{3.5}$$

$$\max_{\boldsymbol{\alpha},\boldsymbol{\beta}} \ D(\boldsymbol{\alpha},\boldsymbol{\beta}) = \sum_{i=1}^{n} \alpha_i - \frac{1}{2} \sum_{i,j=1}^{n} \alpha_i \alpha_j y_i y_j K_{ij} \qquad (3.6)$$

$$-\frac{1}{2\gamma} \sum_{i,j=1}^{n} (\alpha_i + \beta_i - C)(\alpha_j + \beta_j - C) K_{ij}^*$$

$$\text{s.t.} \ \sum_{i=1}^{n} (\alpha_i + \beta_i - C) = 0, \qquad \sum_{i=1}^{n} y_i \alpha_i = 0, \qquad (3.7)$$

$$\forall \ 1 \leq i \leq n, \ \alpha_i \geq 0, \ \beta_i \geq 0. \qquad (3.8)$$

The decision and the correcting functions of SVM+, expressed in the dual variables, are

$$h(\mathbf{x}) = \sum_{j=1}^{n} \alpha_j y_j K(\mathbf{x}_j, \mathbf{x}) + b \qquad (3.9)$$

and $\phi(\mathbf{x}_i^*) = \frac{1}{\gamma} \sum_{j=1}^{n} (\alpha_j + \beta_j - C) K_{ij}^* + d$. Note that SVM and SVM+ have syntactically the same decision function. But semantically they are different, since $\boldsymbol{\alpha}$'s found by SVM and SVM+ can differ a lot.

One of the widely used algorithms for solving (3.3) is SMO, Sequential Minimal Optimization (Platt 1999 [256]). At each iteration SMO optimizes the *working set* of two dual variables, $\alpha_s$ and $\alpha_t$, while keeping all other variables fixed. Note that we cannot optimize the proper subset of such working set, say $\alpha_s$: due to the constraint (3.4), if we fix $n-1$ variables then the last variable is also fixed. Hence, the working sets selected by SMO are *irreducible*.

Following (Bottou & Lin 2007 [38]), we present SMO as an instantiation of the framework of *sparse line search* algorithms to the optimization problem of SVM. At each iteration the algorithms in this framework perform line search in *maximally sparse direction*, which is close to the gradient direction. We instantiate the above framework to SVM+ optimization problem and obtain alternating SMO (aSMO) algorithm. aSMO works with irreducible working sets of two or three variables.

Sparse line search algorithms in turn belong to the family of approximate gradient descent algorithms. The latter algorithms optimize by going roughly in the direction of gradient. Unfortunately (e.g., see Nocedal & Wright 2006 [244]) gradient descent algorithms can have very slow convergence. One of the possible remedies is to use conjugate direction optimization, where each new optimization step is conjugate to all previous ones. In this chapter we combine the ideas of sparse line search and conjugate directions and present a framework of *conjugate sparse line search* algorithms. At each iteration the algorithms in this framework perform line search in a chosen maximally sparse direction which is close to the gradient direction and is conjugate to $k$ previously chosen ones. We instantiate the above framework to SVM+ optimization problem and obtain Conjugate Alternating SMO (caSMO) algorithm. caSMO works with irreducible working sets of size up to $k+2$ and $k+3$, respectively. Our experiments indicate an order-of-magnitude running time improvement of caSMO over aSMO.

Vapnik et al. (Vapnik et al. 2008, 2009 [304, 305]) showed that the LUPI paradigm emerges in several domains, for example, time series prediction and protein classification. To motivate further the usage of the LUPI paradigm and SVM+, we show how it can be used to learn from the data generated by human computation games. Our experiments indicate that the tags generated by the players in ESP (Von Ahn & Dabbish 2004 [307]) and Tag a Tune (Law & Von Ahn 2009 [195]) games result in the improvement in image and music classification.

This chapter has the following structure. In Section 3.2 we present sparse line search optimization framework and instantiate it to SVM and SVM+. In Section 3.3 we present conjugate sparse line search optimization framework and instantiate it to SVM+. In Section 3.4 we prove the convergence of aSMO and caSMO. We present our experimental results in Section 3.5. In Section 3.5.3 we compare aSMO and caSMO with other algorithms for solving SVM+. Finally, Section 3.5.4 describes our experiments with the data generated by human computation games.

**Previous and Related Work**

Blum and Mitchell (Blum & Mitchell 1998 [28]) introduced a semisupervised multiview learning model that leverages the unlabeled examples in order to boost the performance in each of the views. SVM-2K algorithm of Farquhar et al. (Farquhar et al. 2005 [117]), which is similarly to SVM+ is a modification of SVM, performs fully supervised multiview learning without relying on unlabeled examples. The conceptual difference between SVM+ and SVM-2K is that SVM+ looks for a good classifier in a single space, while SVM-2K looks for good classifiers in two spaces. Technically, the optimization problem of SVM+ involves less constraints, dual variables, and hyperparameters than the one of SVM-2K.

Izmailov et al. (Izmailov et al. 2009 [174]) developed SMO-style algorithm, named generalized SMO (gSMO), for solving SVM+. gSMO was used in the experiments in (Vapnik et al. 2008, 2009 [304, 305]). At each iteration gSMO chooses two indices, $i$ and $j$, and optimizes $\alpha_i, \alpha_j, \beta_i, \beta_j$ while fixing other variables. As opposed to the working sets used by aSMO and caSMO, the one used by gSMO is not irreducible. In Section 3.2 we discuss the disadvantages of chosing the working set, which is not irreducible and in Section 3.5.3 we show that aSMO and caSMO are significantly faster than the algorithm of (Izmailov et al. 2009 [174]).

## 3.2    Sparse Line Search Algorithms

We consider the optimization problem $\max_{\mathbf{x} \in \mathcal{F}} f(\mathbf{x})$, where $f$ is a concave function and $\mathcal{F}$ is a convex compact domain.

**Definition 1** *A direction* $\mathbf{u} \in \mathbb{R}^m$ *is* **feasible** *at the point* $\mathbf{x} \in \mathcal{F}$ *if* $\exists \lambda > 0$ *such that* $\mathbf{x} + \lambda \mathbf{u} \in \mathcal{F}$.

**Definition 2** *Let* $NZ(\mathbf{u}) \subseteq \{1, 2, \ldots, n\}$ *be a set of nonzero indices of* $\mathbf{u} \in \mathbb{R}^n$. *A feasible maximally sparse feasible direction* $\mathbf{u}$ *is* **maximally sparse** *if any* $\mathbf{u}' \in \mathbb{R}^n$, *such that* $NZ(\mathbf{u}') \subset NZ(\mathbf{u})$, *is not feasible.*

**Lemma 1** *Let* $\mathbf{u} \in \mathbb{R}^n$ *be a feasible direction vector. The variables in* $NZ(\mathbf{u})$ *have a single degree of freedom iff* $\mathbf{u}$ *is maximally sparse feasible.*

It follows from this lemma that if we choose a maximally sparse feasible direction $\mathbf{u}$ and maximize $f$ along the half-line $\mathbf{x} + \lambda \mathbf{u}$, $\lambda > 0$, then we find the optimal values of the variables in $NZ(\mathbf{u})$ while fixing all other variables. As we will see in the next sections, if $\mathbf{u}$ is sparse and has a constant number of nonzero entries then the exact maximization along the line can be done in a constant time. In the next sections we will also show a fast way of choosing a good maximally sparse feasible direction. To ensure that the optimal $\lambda \neq 0$ we require the direction $\mathbf{u}$ to have an angle with the gradient strictly larger than $\pi/2$. The resulting sparse line search optimization framework is formalized in Algorithm 1. Algorithm 1 has

---

**Algorithm 1** Line search optimization with maximally sparse feasible directions.

---

**Input:** Function $f : \mathbb{R}^n \to \mathbb{R}$, domain $\mathcal{F} \subset \mathbb{R}^n$, initial point $\mathbf{x}_0 \in \mathcal{F}$, constants $\tau > 0$ and $\kappa > 0$.

**Output:** A point $\mathbf{x} \in \mathcal{F}$ with a 'large' value of $f(\mathbf{x})$.

1: Set $\mathbf{x} = \mathbf{x}_0$. Let $\bar{\lambda}(\mathbf{x}, \mathbf{u}) = \max\{\lambda : \mathbf{x} + \lambda\mathbf{u} \in \mathcal{F}\}$.

2: **while** $\exists$ a maximally sparse feasible direction $\mathbf{u}$ such that $\nabla f(\mathbf{x})^T \mathbf{u} > \tau$ and $\bar{\lambda}(\mathbf{x}, \mathbf{u}) > \kappa$ **do**

3:     Choose a maximally sparse feasible direction $\mathbf{u}$ such that $\nabla f(\mathbf{x})^T \mathbf{u} > \tau$ and $\bar{\lambda}(\mathbf{x}, \mathbf{u}) > \kappa$.

4:     Let $\lambda^* = \arg\max_{\lambda : \mathbf{x} + \lambda\mathbf{u} \in \mathcal{F}} f(\mathbf{x} + \lambda\mathbf{u})$.

5:     Set $\mathbf{x} = \mathbf{x} + \lambda^*\mathbf{u}$.

6: **end while**

---

two parameters, $\tau$ and $\kappa$. The parameter $\tau$ controls the minimal angle between the chosen direction $\mathbf{u}$ to the direction of gradient. The parameter $\kappa$ is the minimal size of the optimization step. Algorithm 1 stops when there is no maximally sparse feasible direction that is close to the direction of gradient and allows an optimization step of significant length.

Let $\mathbf{u}$ be a feasible direction that is not maximally sparse. By Definition 2 and Lemma 1, the set $NZ(\mathbf{u})$ has at least two degrees of freedom. Thus, the line search along the half-line $\mathbf{x} + \lambda\mathbf{u}$, $\lambda > 0$, will not find the optimal values of the variables in $NZ(\mathbf{u})$ while fixing all other variables. To achieve the latter, instead of a 1-dimensional line search we need to do the expensive $k$-dimensional hyperplane search, where $k > 1$ is number of the degrees of freedom in $NZ(\mathbf{u})$. This motivates the usage of the maximally sparse search directions.

In Section 3.2.1 we instantiate Algorithm 1 to the optimization problem of SVM and obtain the well-known SMO algorithm. In Section 3.2.2 we instantiate Algorithm 1 to the optimization problem of SVM+.

### 3.2.1    SMO Algorithm for Solving SVM

Following (Bottou & Lin 2007 [38]), we present SMO as a line-search optimization algorithm. Let $I_1(\boldsymbol{\alpha}) = \{i : (\alpha_i > 0 \text{ and } y_i = -1) \text{ or } (\alpha_i < C \text{ and } y_i = 1)\}$, $I_2(\boldsymbol{\alpha}) = \{i : (\alpha_i > 0 \text{ and } y_i = 1) \text{ or } (\alpha_i < C \text{ and } y_i = -1)\}$. At each iteration, SMO chooses a direction $\mathbf{u_s} \in \mathbb{R}^n$ such that $\mathbf{s} = (i, j)$, $i \in I_1$, $j \in I_2$, $u_i = y_i$, $u_j = -y_j$ and for any $r \neq i, j$, $u_r = 0$. If we move from $\boldsymbol{\alpha}$ in the direction $\mathbf{u_s}$ then (3.4) is satisfied. It follows from the definition of $I_1$, $I_2$ and $\mathbf{u_s}$, that $\exists \lambda = \lambda(\boldsymbol{\alpha}, \mathbf{s}) > 0$ such that $\boldsymbol{\alpha} + \lambda\mathbf{u_s} \in [0, C]^n$. Thus, $\mathbf{u_s}$ is a feasible direction. The direction $\mathbf{u_s}$ is also maximally sparse. Indeed, any direction $\mathbf{u}'$ with strictly more zero coordinates than in $\mathbf{u_s}$ has a single nonzero coordinate. But if we move along such $\mathbf{u}'$ then we will not satisfy (3.4).

Let $\mathbf{g}(\boldsymbol{\alpha})$ be a gradient of (3.3) at point $\boldsymbol{\alpha}$. By Taylor expansion of $\psi(\lambda) \triangleq D(\boldsymbol{\alpha}^{\mathrm{old}} + \lambda\mathbf{u_s})$ at $\lambda = 0$, the maximum of $\psi(\lambda)$ along the half-line $\lambda \geq 0$ is at

$$\lambda'(\boldsymbol{\alpha}^{\mathrm{old}}, \mathbf{s}) = -\frac{\frac{\partial D(\boldsymbol{\alpha}^{\mathrm{old}} + \lambda\mathbf{u_s})}{\partial\lambda}\big|_{\lambda=0}}{\frac{\partial^2 D(\boldsymbol{\alpha}^{\mathrm{old}} + \lambda\mathbf{u_s})}{\partial\lambda^2}\big|_{\lambda=0}} = -\frac{\mathbf{g}^T(\boldsymbol{\alpha}^{\mathrm{old}})\mathbf{u_s}}{\mathbf{u_s}^T H \mathbf{u_s}}, \qquad (3.10)$$

where $H$ is a Hessian of $D(\boldsymbol{\alpha}^{\mathrm{old}})$. The clipped value of $\lambda'(\boldsymbol{\alpha}^{\mathrm{old}}, \mathbf{s})$ is

$$\lambda^*(\boldsymbol{\alpha}^{\mathrm{old}}, \mathbf{s}) = \min_{i \in \mathbf{s}} \left( \frac{C - \alpha_i^{\mathrm{old}}}{u_i}, \max_{i \in \mathbf{s}} \left( \lambda'(\boldsymbol{\alpha}^{\mathrm{old}}, \mathbf{s}), -\frac{\alpha_i^{\mathrm{old}}}{u_i} \right) \right) \qquad (3.11)$$

so that the new value $\boldsymbol{\alpha} = \boldsymbol{\alpha}^{\mathrm{old}} + \lambda^*(\boldsymbol{\alpha}^{\mathrm{old}}, \mathbf{s})\mathbf{u_s}$ is always in the domain $[0, C]^n$.

Let $\tau > 0$ be a small constant and $I = \{\mathbf{u_s} \mid \mathbf{s} = (i,j),\ i \in I_1,\ j \in I_2,\ \mathbf{g}(\boldsymbol{\alpha}^{\text{old}})^T \mathbf{u_s} > \tau\}$. If $I = \emptyset$ then SMO stops. It can be shown that in this case Karush-Kuhn-Tucker (KKT) optimality conditions are almost satisfied (up to accuracy $\tau$).

Let $I \neq \emptyset$. We now describe a procedure for choosing the next direction. This procedure is currently implemented in Library for Support Vector Machines (LIBSVM) and has complexity of $O(n)$. Let

$$\mathbf{s} = \arg \max_{\mathbf{t}: \mathbf{u_t} \in I} \mathbf{g}(\boldsymbol{\alpha}^{\text{old}})^T \mathbf{u_t} \qquad (3.12)$$

be the nonzero components of the direction $\mathbf{u}$ that is maximally aligned with the gradient. Earlier versions of LIBSVM chose $\mathbf{u_s}$ as the next direction. As observed by (Fan et al. 2005 [116]), empirically faster convergence is obtained if instead of $\mathbf{s} = (i,j)$ we choose $\mathbf{s}' = (i,j')$ such that $\mathbf{u_{s'}} \in I$ and the move in the direction $\mathbf{u_{s'}}$ maximally increases (3.3):

$$\mathbf{s}' = \arg \max_{\substack{\mathbf{t}=(t_1,t_2) \\ t_1=i, \mathbf{u_t} \in I}} D(\boldsymbol{\alpha}^{\text{old}} + \lambda'(\boldsymbol{\alpha}^{\text{old}}, \mathbf{t})\mathbf{u_t}) - D(\boldsymbol{\alpha}) \qquad (3.13)$$

$$= \arg \max_{\substack{\mathbf{t}=(t_1,t_2) \\ t_1=i, \mathbf{u_t} \in I}} \frac{(\mathbf{g}(\boldsymbol{\alpha}^{\text{old}})^T \mathbf{u_t})^2}{\mathbf{u_t}^T H \mathbf{u_t}}. \qquad (3.14)$$

The equality (3.14) follows from substituting $\lambda'(\boldsymbol{\alpha}^{\text{old}}, \mathbf{t})$ (given by (3.10)) into (3.13) and using the definition (3.3) of $D(\boldsymbol{\alpha})$.

### 3.2.2  Alternating SMO (aSMO) for Solving SVM+

We start with the identification of the maximally sparse feasible directions. Since (3.6) has $2n$ variables, $\{\alpha_i\}_{i=1}^n$ and $\{\beta_i\}_{i=1}^n$, the search direction $\mathbf{u} \in \mathbb{R}^{2n}$. We designate the first $n$ coordinates of $\mathbf{u}$ for $\boldsymbol{\alpha}$'s and the last $n$ coordinates for $\boldsymbol{\beta}$'s. It follows from the equality constraints (3.7) that (3.6) has three types of sets of maximally sparse feasible directions, $I_1, I_2, I_3 \subset \mathbb{R}^{2n}$:

1. $I_1 = \{\mathbf{u_s} \mid \mathbf{s} = (s_1, s_2),\ n+1 \leq s_1, s_2 \leq 2n,\ s_1 \neq s_2;\ u_{s_1} = 1, u_{s_2} = -1, \beta_{s_2} > 0,\ \forall\, i \notin \mathbf{s}\ u_i = 0\}$. A move in the direction $\mathbf{u_s} \in I_1$ increases $\beta_{s_1}$ and decreases $\beta_{s_2} > 0$ by the same quantity. The rest of the variables remain fixed.

2. $I_2 = \{\mathbf{u_s} \mid \mathbf{s} = (s_1, s_2),\ 0 \leq s_1, s_2 \leq n,\ s_1 \neq s_2,\ y_{s_1} = y_{s_2};\ u_{s_1} = 1, u_{s_2} = -1, \alpha_{s_2} > 0,\ \forall\, i \notin \mathbf{s}\ u_i = 0\}$. A move in the direction $\mathbf{u_s} \in I_2$ increases $\alpha_{s_1}$ and decreases $\alpha_{s_2} > 0$ by the same quantity. The rest of the variables remain fixed.

3. $I_3 = \{\mathbf{u_s} \mid \mathbf{s} = (s_1, s_2, s_3),\ 0 \leq s_1, s_2 \leq n,\ n+1 \leq s_3 \leq 2n,\ s_1 \neq s_2;\ \forall\, i \notin \mathbf{s}\ u_i = 0;\ y_{s_1} \neq y_{s_2};\ u_{s_1} = u_{s_2} = 1, u_{s_3} = -2, \beta_{s_3} > 0$ or $u_{s_1} = u_{s_2} = -1, u_{s_3} = 2, \alpha_{s_1} > 0, \alpha_{s_2} > 0\}$. A move in the direction $\mathbf{u_s} \in I_3$ either increases $\alpha_{s_1}$, $\alpha_{s_2}$ and decreases $\beta_{s_3} > 0$, or decreases $\alpha_{s_1} > 0$, $\alpha_{s_2} > 0$ and increases $\beta_{s_3}$. The absolute value of the change in $\beta_{s_3}$ is twice the absolute value of the change in $\alpha_{s_1}$ and in $\alpha_{s_2}$. The rest of the variables remain fixed.

We abbreviate $\mathbf{x} = (\boldsymbol{\alpha}, \boldsymbol{\beta})^T$ and $\mathbf{x}^{\text{old}} = (\boldsymbol{\alpha}^{\text{old}}, \boldsymbol{\beta}^{\text{old}})^T$. It can be verified that if we move from any feasible point $\mathbf{x}$ in the direction $\mathbf{u_s} \in I_1 \cup I_2 \cup I_3$ then (3.7) is satisfied. Let $D(\mathbf{x})$ be an objective function defined by (3.6) and $\lambda^*(\mathbf{x}^{\text{old}}, \mathbf{s})$ be a real number that maximizes $\psi(\lambda) = D(\mathbf{x}^{\text{old}} + \lambda \mathbf{u_s})$ such that $\mathbf{x}^{\text{old}} + \lambda \mathbf{u_s}$ satisfies (3.8). The optimization step in aSMO is $\mathbf{x} = \mathbf{x}^{\text{old}} + \lambda^*(\mathbf{x}^{\text{old}}, \mathbf{s}) \cdot \mathbf{u_s}$, where $\mathbf{u_s} \in I_1 \cup I_2 \cup I_3$. We now show how $\lambda^*(\mathbf{x}^{\text{old}}, \mathbf{s})$ is computed.

Let $\mathbf{g}(\mathbf{x}^{\text{old}})$ and $H$ be respectively gradient and Hessian of (3.6) at the point $\mathbf{x}^{\text{old}}$. Using the Taylor expansion of $\psi(\lambda)$ around $\lambda = 0$ we obtain that

$$\lambda'(\mathbf{x}^{\text{old}}, \mathbf{s}) = \arg \max_{\lambda: \lambda \geq 0} \psi(\lambda) = -\frac{\frac{\partial D(\mathbf{x}^{\text{old}} + \lambda \mathbf{u_s})}{\partial \lambda}\big|_{\lambda=0}}{\frac{\partial^2 D(\mathbf{x}^{\text{old}} + \lambda \mathbf{u_s})}{\partial \lambda^2}\big|_{\lambda=0}} = -\frac{\mathbf{g}(\mathbf{x}^{\text{old}})^T \mathbf{u_s}}{\mathbf{u_s}^T H \mathbf{u_s}}. \qquad (3.15)$$

---

**Algorithm 2** Conjugate line search optimization with maximally sparse feasible directions.

---

**Input:** Function $f(\mathbf{x}) = \frac{1}{2}\mathbf{x}^T Q \mathbf{x} + \mathbf{c}^T \mathbf{x}$, domain $\mathcal{F} \subset \mathbb{R}^n$, initial point $\mathbf{x}_0 \in \mathcal{F}$, constants $\tau > 0$ and $\kappa > 0$, maximal order of conjugacy $k_{\max} \geq 1$.

**Output:** A point $\mathbf{x} \in \mathcal{F}$ with 'large' value of $f(\mathbf{x})$.

1: Set $\mathbf{x} = \mathbf{x}_0$. Let $\bar{\lambda}(\mathbf{x}, \mathbf{u}) = \max\{\lambda \; : \; \mathbf{x} + \lambda\mathbf{u} \in \mathcal{F}\}$.

2: **while** $\exists$ a maximally sparse feasible direction $\mathbf{u}$ such that $\bigtriangledown f(\mathbf{x})^T\mathbf{u} > \tau$ and $\bar{\lambda}(\mathbf{x}, \mathbf{u}) > \kappa$ **do**

3:     Choose an order of conjugacy $k$, $0 \leq k \leq k_{\max}$.

4:     Let $\mathbf{g}_0$ and $\mathbf{g}_i$ be the gradient of $f$ at the beginning of current and the $i$-th previous iteration.

5:     Choose a maximally sparse feasible direction $\mathbf{u}$ such that $\bigtriangledown f(\mathbf{x})^T\mathbf{u} > \tau$, $\bar{\lambda}(\mathbf{x}, \mathbf{u}) > \kappa$ and for all $1 \leq i \leq k$, $\mathbf{u}^T(\mathbf{g}_{i-1} - \mathbf{g}_i) = 0$.

6:     Let $\lambda^* = \arg\max_{\lambda:\mathbf{x}+\lambda\mathbf{u}\in\mathcal{F}} f(\mathbf{x} + \lambda\mathbf{u})$.

7:     Set $\mathbf{x} = \mathbf{x} + \lambda^*\mathbf{u}$.

8: **end while**

---

We clip the value of $\lambda'(\mathbf{x}^{\mathrm{old}}, \mathbf{s})$ so that the new point $\mathbf{x} = \mathbf{x}^{\mathrm{old}} + \lambda^*(\mathbf{x}_{\mathrm{old}}, \mathbf{s}) \cdot \mathbf{u_s}$ will satisfy (3.8):

$$\lambda^*(\mathbf{x}^{\mathrm{old}}, \mathbf{s}) = \max_{i \in \mathbf{s}}\left(\lambda'(\mathbf{x}_{\mathrm{old}}, \mathbf{s}), -\frac{x_i^{\mathrm{old}}}{u_i}\right). \tag{3.16}$$

Let $\tau > 0$, $\kappa > 0$, $\bar{\lambda}(\mathbf{x}^{\mathrm{old}}, \mathbf{s}) = \max\{\lambda | \mathbf{x}^{\mathrm{old}} + \lambda\mathbf{u_s} \text{ satisfies (3.8)}\}$ and $I = \{\mathbf{u_s} \mid \mathbf{u_s} \in I_1 \cup I_2 \cup I_3, \mathbf{g}(\mathbf{x}^{\mathrm{old}})^T\mathbf{u_s} > \tau \text{ and } \bar{\lambda}(\mathbf{x}^{\mathrm{old}}, \mathbf{s}) > \kappa\}$. If $I = \emptyset$ then aSMO stops.

Suppose that $I \neq \emptyset$. We choose the next direction in three steps. The first two steps are similar to the ones done by LIBSVM (Chang & Lin 2001 [67]). At the first step, for each $1 \leq i \leq 3$ we find a vector $\mathbf{u}_{\mathbf{s}^{(i)}} \in I_i$ that has a minimal angle with $\mathbf{g}(\mathbf{x}^{\mathrm{old}})$ among all vectors in $I_i$: if $I_i \neq \emptyset$ then $\mathbf{s}^{(i)} = \arg\max_{\mathbf{s}:\mathbf{u_s}\in I_i} \mathbf{g}(\mathbf{x}^{\mathrm{old}})^T\mathbf{u_s}$, otherwise $\mathbf{s}^{(i)}$ is empty.

For $1 \leq i \leq 3$, let $\widetilde{I}_i = \{\mathbf{u_s} \mid \mathbf{u_s} \in I_i, \mathbf{g}(\mathbf{x}^{\mathrm{old}})^T\mathbf{u_s} > \tau \text{ and } \bar{\lambda}(\mathbf{x}^{\mathrm{old}}, \mathbf{u_s}) > \kappa\}$. At the second step for $1 \leq i \leq 2$, if the pair $\mathbf{s}^{(i)} = (s_1^{(i)}, s_2^{(i)})$ is not empty then we fix $s_1^{(i)}$ and find $\mathbf{s}'^{(i)} = (s_1^{(i)}, s_2'^{(i)})$ such that $\mathbf{u}_{\mathbf{s}'^{(i)}} \in \widetilde{I}_i$ and the move in the direction $\mathbf{u}_{\mathbf{s}'^{(i)}}$ maximally increases (3.6):

$$\begin{aligned}
\mathbf{s}'^{(i)} &= \arg\max_{\substack{\mathbf{t}:\mathbf{t}=(t_1,t_2)\\ t_1=s_1^{(i)}, \mathbf{u_t}\in\widetilde{I}_i}} \underbrace{\frac{D(\mathbf{x}^{\mathrm{old}} + \lambda'(\mathbf{x}^{\mathrm{old}}, \mathbf{t}) \cdot \mathbf{u_t}) - D(\mathbf{x}^{\mathrm{old}})}{}}_{\triangleq \Delta_i(\mathbf{x}^{\mathrm{old}}, \mathbf{t})} \\
&= \arg\max_{\substack{\mathbf{t}:\mathbf{t}=(t_1,t_2)\\ t_1=s_1^{(i)}, \mathbf{u_t}\in\widetilde{I}_i}} \frac{(\mathbf{g}(\mathbf{x}^{\mathrm{old}})^T\mathbf{u_t})^2}{\mathbf{u_t}^T H \mathbf{u_t}}.
\end{aligned} \tag{3.17}$$

The right-hand equality in (3.17) follows from substituting $\lambda'(\mathbf{x}^{\mathrm{old}}, \mathbf{t})$ (given by (3.15)) into $\Delta_i(\mathbf{x}^{\mathrm{old}}, \mathbf{t})$ and using the definition (3.6) of $D(\mathbf{x})$. Similarly, for $i = 3$ if the triple $\mathbf{s}^{(3)} = (s_1^{(3)}, s_2^{(3)}, s_3^{(3)})$ is not empty then we fix $s_1^{(3)}$ and $s_3^{(3)}$ and find $\mathbf{s}'^{(3)} = (s_1^{(3)}, s_2'^{(3)}, s_3^{(3)})$ such that $\mathbf{u}_{\mathbf{s}'^{(3)}} \in \widetilde{I}_3$ and the move in the direction $\mathbf{u}_{\mathbf{s}'^{(3)}}$ maximally increases (3.6). At the third step, we choose $j = \arg\max_{1\leq i\leq 3, \mathbf{s}'^{(i)}\neq\emptyset} \Delta_i(\mathbf{x}^{\mathrm{old}}, \mathbf{s}'^{(i)})$. The final search direction is $\mathbf{u}_{\mathbf{s}'^{(j)}}$.

Since the direction vectors $\mathbf{u_s} \in I_1 \cup I_2 \cup I_3$ have a constant number of nonzero components, it can be shown that our procedure for choosing the next direction has a complexity of $O(n)$. The following theorem, proven in Section 3.4, establishes the asymptotic convergence of aSMO.

**Theorem 1** *aSMO stops after a finite number of iterations. Let* $\mathbf{x}^*(\tau, \kappa)$ *be the solution found by aSMO for a given* $\tau$ *and* $\kappa$ *and let* $\mathbf{x}^*$ *be a solution of (3.6). Then* $\lim_{\tau \to 0, \kappa \to 0} \mathbf{x}^*(\tau, \kappa) = \mathbf{x}^*$.

---

## 3.3 Conjugate Sparse Line Search

Conjugate sparse line search is an extension of Algorithm 1. Similarly to Algorithm 1, we choose a maximally sparse feasible direction. However, in contrast to Algorithm 1, in a conjugate sparse line search we also require that the next search direction is conjugate to $k$ previously chosen ones. The order of conjugacy, $k$, is between 0 and an upper bound $k_{\max} \geq 1$. We allow $k$ to vary between iterations. Note that if $k = 0$ then the optimization step done by conjugate sparse line search is the same as the one done by a (nonconjugate) sparse line search. In Section 3.3.1 we show how we choose $k$.

Let $Q$ be $n \times n$ positive semidefinite matrix and $\mathbf{c} \in \mathbb{R}^n$. In this section we assume that $f$ is a quadratic function, $f(\mathbf{x}) = \frac{1}{2}\mathbf{x}^T Q \mathbf{x} + \mathbf{c}^T \mathbf{x}$. This assumption is satisfied by all SVM-like problems. Let $\mathbf{u}$ be the next search direction that we want to find, and $\mathbf{u}_i$ be a search direction found at the $i$-th previous iteration. We would like $\mathbf{u}$ to be conjugate to $\mathbf{u}_i$, $1 \leq i \leq k$, w.r.t. the matrix $Q$:

$$\forall 1 \leq i \leq k, \ \mathbf{u}^T Q \mathbf{u}_i = 0 \ . \tag{3.18}$$

Let $\mathbf{g}_0$ and $\mathbf{g}_i$ be the gradient of $f$ at the beginning of current and the $i$-th previous iteration, respectively. Similarly, let $\mathbf{x}_0$ and $\mathbf{x}_i$ be the optimization point at the beginning of current and the $i$-th previous iteration. By the definition of $f$, $\mathbf{g}_i = Q\mathbf{x}_i + \mathbf{c}$. Since for any $1 \leq i \leq n$, $\mathbf{x}_{i-1} = \mathbf{x}_i + \lambda_i \mathbf{u}_i$, where $\lambda_i \in \mathbb{R}$, we have that $\mathbf{u}^T Q \mathbf{u}_i = \mathbf{u}^T Q(\mathbf{x}_{i-1} - \mathbf{x}_i)/\lambda_i = \mathbf{u}^T(\mathbf{g}_{i-1} - \mathbf{g}_i)/\lambda_i$, and thus the conjugacy condition (3.18) is equivalent to

$$\forall 1 \leq i \leq k, \ \mathbf{u}^T(\mathbf{g}_{i-1} - \mathbf{g}_i) = 0 \ . \tag{3.19}$$

The resulting conjugate sparse line search algorithm is formalized in Algorithm 2.

Suppose that the search directions $\mathbf{u}$ and $\mathbf{u}_i$, $1 \leq i \leq k$, have $r$ nonzero components. Commonly the implementations of SMO-like algorithms (e.g., LIBSVM and UniverSVM) compute the gradient vector after each iteration. Thus, the additional complexity of checking the condition (3.19) is $O(kr)$. However, the additional complexity of checking (3.18) is $O(kr^2)$. As we will see in Section 3.3.1, if $k \geq 1$ then $r \geq 3$. Hence, it is faster to check (3.19) than (3.18).

In the rest of this section we describe the instantiation of Algorithm 2 to the optimization problem of SVM+.

### 3.3.1 Conjugate Alternating SMO (caSMO) for Solving SVM+

We use the same notation as in Section 3.2.2. At each iteration we chose a maximally sparse feasible direction $\mathbf{u_s} \in \mathbb{R}^{2n}$ that solves the following system of $k + 2$ linear equalities:

$$\begin{cases} \mathbf{u_s}^T \bar{\mathbf{y}} = 0, \quad \mathbf{u_s}^T \mathbf{1} = 0 \\ \mathbf{u_s}^T(\mathbf{g}_{i-1} - \mathbf{g}_i) = 0 \quad \forall 1 \leq i \leq k \end{cases} \tag{3.20}$$

where $\mathbf{1} \in \mathbb{R}^{2n}$ is a column vector with all entries being one and $\bar{\mathbf{y}} = [\mathbf{y}; \mathbf{0}]$ is a row-wise concatenation of the label vector $\mathbf{y}$ with $n$-dimensional zero column vector. It can be verified that if we move from any feasible point $\mathbf{x}$ in the direction $\mathbf{u_s}$ defined above then (3.7) is satisfied.

We assume that the vectors $\overline{\mathbf{y}}, \mathbf{1}, \mathbf{g}_0 - \mathbf{g}_1, \ldots, \mathbf{g}_{k-1} - \mathbf{g}_k$ are linearly independent. In all our experiments this assumption was indeed satisfied. It follows from the equality constraints (3.7) that there are 4 types of maximally sparse feasible directions $\mathbf{u_s}$ that solve (3.20):

1. $J_1 = \{\mathbf{u_s} \mid \mathbf{u_s}$ is feasible, $\mathbf{s} = (s_1, \ldots, s_{k+2}), \; n+1 \leq s_1, \ldots, s_{k+2} \leq 2n; \forall \, i \notin \mathbf{s} \; u_i = 0\}$. A move in the direction $\mathbf{u_s} \in I_1$ changes $\beta$'s indexed by $\mathbf{s}$. The rest of the variables are fixed.

2. $J_2 = \{\mathbf{u_s} \mid \mathbf{u_s}$ is feasible, $\mathbf{s} = (s_1, \ldots, s_{k+2}), \; 1 \leq s_1, \ldots, s_{k+2} \leq n, \; y_{s_1} = \cdots = y_{s_{k+2}}; \; \forall \, i \notin \mathbf{s} \; u_i = 0\}$. A move in the direction $\mathbf{u_s} \in I_2$ changes $\alpha$'s indexed by $\mathbf{s}$ such that the corresponding examples have the same label. The rest of the variables remain fixed.

3. $J_3 = \{\mathbf{u_s} \mid \mathbf{u_s}$ is feasible, $\mathbf{s} = (s_1, \ldots, s_{k+3}); \; \exists \, i, j$ such that $s_i, s_j \leq n$ and $y_{s_i} \neq y_{s_j}; \; \exists \, r$ such that $s_r > n; \; \forall \, i \notin \mathbf{s}, \; u_i = 0\}$. A move in the direction $\mathbf{u_s} \in I_3$ changes at least one $\beta$ and at least $\alpha$'s such that the corresponding two examples have different labels. The variables that are not indexed by $\mathbf{s}$ remain fixed.

4. $J_4 = \{\mathbf{u_s} \mid \mathbf{u_s}$ is feasible, $\mathbf{s} = (s_1, \ldots, s_{k+3}), \; k \geq 1, \; 1 \leq s_1, \ldots, s_{k+3} \leq n; \; \exists \, i \neq j$ such that $y_{s_i} = y_{s_j} = 1; \; \exists \, r \neq t$ such that $y_{s_r} = y_{s_t} = -1; \; \forall \, i \notin \mathbf{s}, \; u_i = 0\}$. A move in the direction $\mathbf{u_s} \in I_2$ changes $\alpha$'s indexed by $\mathbf{s}$. At least two such $\alpha$'s correspond to different positive examples and at least two such $\alpha$'s correspond to different negative examples. The rest of the variables remain fixed.

The definitions of $J_1$-$J_3$ here are direct extensions of the corresponding definitions of $I_1$-$I_3$ for aSMO (see Section 3.2.2). However, the definition of $J_4$ does not have a counterpart in aSMO. By the definition of $J_4$, it contains feasible direction vectors that change at least two $\alpha$'s of positive examples and at least two $\alpha$'s of negative examples. But such vectors are not maximally sparse feasible for the original optimization problem (3.6), since for any $\mathbf{u}_{s'} \in J_4$ there exists $\mathbf{u}_{s'} \in I_2$ such that $NZ(\mathbf{u}_{s'}) \subset NZ(\mathbf{u_s})$. Hence, the vectors from $J_4$ cannot be used by aSMO. Thus, the conjugate direction constraints give additional power to the optimization algorithm by allowing to use a new type of directions that were not available previously.

The optimization step in caSMO is $\mathbf{x} = \mathbf{x}^{\text{old}} + \lambda^*(\mathbf{x}^{\text{old}}, \mathbf{s}) \cdot \mathbf{u_s}$, where $\mathbf{u_s}$ satisfies the above conditions and $\lambda^*(\mathbf{x}^{\text{old}}, \mathbf{s})$ is computed exactly as in aSMO, by (3.15) and (3.16).

We now describe the procedure for choosing the next direction $\mathbf{u_s}$ that satisfies (3.20). At the first iteration we choose the direction $\mathbf{u_s} \in I_1 \cup I_2 \cup I_3$ that is generated by aSMO and set $k = 0$. At the next iterations we proceed as follows. Let $\mathbf{u_s}$ be the direction chosen at the previous iteration and $\mathbf{s} = (s_1, \ldots, s_{k'})$ be the current working set. If $\mathbf{u_s} \in I_1 \cup I_2$ then $k' = k+2$, otherwise $k' = k+3$. Recall that $0 \leq k \leq k_{\max}$. If $k < k_{\max}$ then we try to increase the order of conjugacy by setting $k = k+1$ and finding an index

$$s_{k'+1} = \arg \max_{\substack{t:\mathbf{s} = (s_1, \ldots, s_{k'}, s_t), \\ \mathbf{u_s} \in J_1 \cup J_2 \cup J_3 \cup J_4 \\ \text{and solves (3.20)}}} \underbrace{D(\mathbf{x}^{\text{old}} + \lambda'(\mathbf{x}^{\text{old}}, \mathbf{s}) \cdot \mathbf{u_s}) - D(\mathbf{x}^{\text{old}})}_{\triangleq \Delta(\mathbf{x}^{\text{old}}, \mathbf{s})}. \tag{3.21}$$

Similarly to (3.17), it can be shown that $\Delta(\mathbf{x}^{\text{old}}, \mathbf{s}) = \frac{(\mathbf{g}(\mathbf{x}^{\text{old}})^T \mathbf{u_s})^2}{\mathbf{u_s}^T H \mathbf{u_s}}$.

If $k = k_{\max}$ then we reduce ourselves to the case of $k < k_{\max}$ and proceed as above. The reduction is done by removing from $\mathbf{s}$ the least recently added index $s_1$, obtaining $\mathbf{s} = (s_2, \ldots, s_{k'})$, downgrading $k = k-1$ and renumerating the indices so that $\mathbf{s} = (s_1, \ldots, s_{k'-1})$.

We also use a "second opinion" and consider the search direction $\mathbf{u}_{s'} \in I_1 \cup I_2 \cup I_3$ that is found by aSMO. If $s_{k'+1}$ found by (3.21) is empty, namely there is no index $t$ such that $\mathbf{s} = (s_1, \ldots, s_{k'}, t)$, $\mathbf{u_s} \in J_1 \cup J_2 \cup J_3 \cup J_4$ and $\mathbf{u_s}$ solves (3.20), then we set $k = 0$ and the next search direction is $\mathbf{u}_{s'}$. Also, if $s_{k'+1}$ is not empty and the functional gain from $\mathbf{u}_{s'}$ is

---

**Algorithm 3** Optimization procedure of (Bordes et al. 2005 [31]).

---

**Input:** Function $f : \mathbb{R}^n \to \mathbb{R}$, domain $\mathcal{F} \subset \mathbb{R}^n$, set of directions $\mathcal{U} \subseteq \mathbb{R}^n$, initial point
$\quad$ $\mathbf{x}_0 \in \mathcal{F}$, constants $\tau > 0$, $\kappa > 0$.

1: Set $\mathbf{x} = \mathbf{x}_0$. Let $\bar{\lambda}(\mathbf{x}, \mathbf{u}) = \max\{\lambda \; : \; \mathbf{x} + \lambda \mathbf{u} \in \mathcal{F}\}$.
2: **while** exists $\mathbf{u} \in \mathcal{U}$ such that $\nabla f(\mathbf{x})^T \mathbf{u} > \tau$ and $\bar{\lambda}(\mathbf{x}, \mathbf{u}) > \kappa$ **do**
3: $\quad$ Choose any $\mathbf{u} \in \mathcal{U}$ such that $\nabla f(\mathbf{x})^T \mathbf{u} > \tau$ and $\bar{\lambda}(\mathbf{x}, \mathbf{u}) > \kappa$.
4: $\quad$ Let $\lambda^* = \arg \max_{\lambda : \mathbf{x} + \lambda \mathbf{u} \in \mathcal{F}} f(\mathbf{x} + \lambda \mathbf{u})$.
5: $\quad$ Set $\mathbf{x} = \mathbf{x} + \lambda^* \mathbf{u}$.
6: **end while**
7: Output $\mathbf{x}$.

---

larger than the one from $\mathbf{u_s}$, namely if $\Delta(\mathbf{x}^{\text{old}}, \mathbf{s}') > \Delta(\mathbf{x}^{\text{old}}, \mathbf{s})$, then we set $k = 0$ and the
next search direction is $\mathbf{u_{s'}}$. Otherwise the next search direction is $\mathbf{u_s}$.

$\quad$ Given the indices $\mathbf{s} = (s_1, \ldots, s_{k'})$, in order to find $\mathbf{u_s}$ we need to solve (3.20). Since $\mathbf{u_s}$
has $O(k)$ nonzero entries, this takes $O(k^3)$ time. It can be shown that the overall complexity
of the above procedure of choosing the next search direction is $O(k_{\max}^3 n)$. Hence, it is
computationally feasible only for small values of $k_{\max}$. In our experiments (Section 3.5.3)
we observed that the value of $k_{\max}$ that gives nearly-optimal running time is 3. The following
theorem, proven in Section 3.4, establishes the asymptotic convergence of caSMO.

**Theorem 2** *caSMO stops after a finite number of iterations. Let* $\mathbf{x}^*(\tau, \kappa)$ *be the solution found by caSMO for a given* $\tau$ *and* $\kappa$ *and let* $\mathbf{x}^*$ *be a solution of (3.6). Then*
$\lim_{\tau \to 0, \kappa \to 0} \mathbf{x}^*(\tau, \kappa) = \mathbf{x}^*$.

---

## 3.4   Proof of Convergence Properties of aSMO, caSMO

We use the results of (Bordes et al. 2005 [31]) to demonstrate convergence properties of aSMO
and caSMO. Bordes et al. (Bordes et al. 2005 [31]) presented a generalization of Algorithm 1.
Instead of maximally sparse vectors $\mathbf{u}$, at each iteration they consider directions $\mathbf{u}$ from
a set $\mathcal{U} \subset \mathbb{R}^n$. The optimization algorithm of (Bordes et al. 2005 [31]) is formalized in
Algorithm 3. aSMO is obtained from Algorithm 3 by setting $\mathcal{U} = I_1 \cup I_2 \cup I_3$.

$\quad$ Our results are based on the following definition and theorem of (Bordes et al. 2005 [31]).

**Definition 3 ((Bordes et al. 2005 [31]))** *A set of directions* $\mathcal{U} \subset \mathbb{R}^n$ *is a **finite wit-
ness family** for a convex set* $\mathcal{F}$ *if* $\mathcal{U}$ *is finite and for any* $\mathbf{x} \in \mathcal{F}$, *any feasible direction* $\mathbf{u}$ *at
the point* $\mathbf{x}$ *is a positive linear combination of a finite number of feasible directions* $\mathbf{v}_i \in \mathcal{U}$
*at the point* $\mathbf{x}$.

The following theorem is a straightforward generalization of Proposition 13 in (Bordes et
al. 2005 [31]).

**Theorem 3 ((Bordes et al. 2005 [31]))** *Let* $\mathcal{U} \subset \mathbb{R}^n$ *be a set that contains a finite witness family for* $\mathcal{F}$. *Then Algorithm 3 stops after a finite number of iterations. Let* $\mathbf{x}^*(\tau, \kappa)$
*be the solution found by Algorithm 3 for a given* $\tau$ *and* $\kappa$ *and let* $\mathbf{x}^* = \arg \min_{\mathbf{x} \in \mathcal{F}} f(\mathbf{x})$.
*Then* $\lim_{\tau \to 0, \kappa \to 0} \mathbf{x}^*(\tau, \kappa) = \mathbf{x}^*$.

**Lemma 2** *The set* $I_1 \cup I_2 \cup I_3$ *(defined in Section 3.2.2) is a finite witness family for the
set* $\mathcal{F}$ *defined by (3.7)-(3.8).*

**Proof** We use the ideas of the proof of Proposition 7 in (Bordes et al. 2005 [31]) showing that $\mathcal{U} = \{\mathbf{e}_i - \mathbf{e}_j, i \neq j\}$ is a finite witness family for SVM. The set $I_1 \cup I_2 \cup I_3$ is finite. Let $\mathbf{x} \in \mathbb{R}^{2n}$ be an arbitrary point in $\mathcal{F}$ and $\mathbf{u}$ be a feasible direction at $\mathbf{x}$. Recall that the first $n$ coordinates of $\mathbf{x}$ are designated for the values of $\boldsymbol{\alpha}$'s and the last $n$ coordinates of $\mathbf{x}$ are designated for the values of $\boldsymbol{\beta}$'s. There exists $\lambda > 0$ such that $\mathbf{x}' = \mathbf{x} + \lambda\mathbf{u} \in \mathcal{F}$. We construct a finite path $\mathbf{x} = \mathbf{z}(0) \to \mathbf{z}(1) \to \ldots \to \mathbf{z}(m-1) \to \mathbf{z}(m) = \mathbf{x}'$ such that $\mathbf{z}(j+1) = \mathbf{z}(j) + \gamma(j)\mathbf{v}(j)$, where $\mathbf{z}(j+1) \in \mathcal{F}$ and $\mathbf{v}(j)$ is a feasible direction at $\mathbf{x}$. We prove by induction on $m$ that these two conditions are satisfied. As an auxiliary tool, we also prove by induction on $m$ that our construction satisfies the following invariant for all $1 \leq i \leq 2n$ and $1 \leq j \leq m$:

$$x_i > x_i' \Leftrightarrow z_i(j) \geq x_i' . \tag{3.22}$$

Let $I_+ = \{i \mid 1 \leq i \leq n,\ y_i = +1\}$ and $I_- = \{i \mid 1 \leq i \leq n,\ y_i = -1\}$. We have from (3.7) that

$$\sum_{i \in I_+} x_i' = \sum_{i \in I_-} x_i' . \tag{3.23}$$

The construction of the path has three steps. At the first step (Cases 1 and 2 below), we move the current overall weight of $\boldsymbol{\beta}$'s (which is $\sum_{i=n+1}^{2n} z_i(j)$) towards the overall weight of $\boldsymbol{\beta}$'s in $\mathbf{x}'$ (which is $\sum_{i=n+1}^{2n} \mathbf{x}'(j)$). When the current overall weight of $\boldsymbol{\beta}$'s is the same as the overall weight of $\boldsymbol{\beta}$'s in $\mathbf{x}'$, we proceed to the second step. At the second step (Case 3 below), we move the values of $\boldsymbol{\beta}$'s in $\mathbf{z}(j)$ towards their corresponding target values at $\mathbf{x}'$. The second step terminates when the values of $\boldsymbol{\beta}$'s in $\mathbf{z}(j)$ and $\mathbf{x}'$ are exactly the same. At the third step (Case 4 below), we move the values of $\boldsymbol{\alpha}$'s in $\mathbf{z}(j)$ towards their corresponding target values at $\mathbf{x}'$. The third step terminates when the values of $\boldsymbol{\beta}$'s in $\mathbf{z}(j)$ and $\mathbf{x}'$ are the same. We now describe this procedure in the formal way.

For each $0 \leq j \leq m-1$, we have four cases:

**Case 1** $\sum_{i=n+1}^{2n} z_i(j) > \sum_{i=n+1}^{2n} x_i'$.
Since $\mathbf{z}(j), \mathbf{x}' \in \mathcal{F}$, by (3.7) we have that $\sum_{i=1}^n z_i(j) < \sum_{i=1}^n x_i'$. Let $T = \{i \mid n+1 \leq i \leq 2n, z_i(j) > x_i'\}$. We have from (3.7) that $\sum_{i \in I_+} z_i(j) = \sum_{i \in I_-} z_i(j)$. Using (3.23) we obtain that there exist $r \in I_+$ and $s \in I_-$ such that $z_r(j) < x_r'$ and $z_s(j) < x_s'$. Let $t$ be any index from $T$, $\mathbf{u}_{rst} \in I_3$ such that $u_t = -2$ and

$$\lambda = \min\left(x_r' - z_r(j), x_s' - z_s(j), (z_t(j) - x_t')/2\right) .$$

Note that $\lambda > 0$. We set $\mathbf{v}(j) \triangleq \mathbf{u}_{rst}$, $\gamma(j) = \lambda$ and $\mathbf{z}(j+1) = \mathbf{z}(j) + \gamma(j)\mathbf{v}(j)$. Since $\mathbf{z}(j) \in \mathcal{F}$, it follows from the definition of $\lambda$ and $\mathbf{u}_{rst}$ that $\mathbf{z}(j+1) \in \mathcal{F}$. Since $\mathbf{z}(j)$ satisfies (3.22), we have that $x_t > x_t' \geq 0$. Thus $\mathbf{u}_{rst}$ is a feasible direction at $\mathbf{x}$. Finally, it follows from the definition of $r$, $s$, $t$, $\lambda$ and $\mathbf{u}_{rst}$ that $z_i(j+1) \geq x_i' \Leftrightarrow z_i(j) \geq x_i'$ for any $1 \leq i \leq 2n$. Hence, $\mathbf{z}(j+1)$ satisfies (3.22).

**Case 2** $\sum_{i=n+1}^{2n} z_i(j) < \sum_{i=n+1}^{2n} x_i'$.
The treatment of this case is similar to what was done with the previous one. Since $\mathbf{z}(j), \mathbf{x}' \in \mathcal{F}$, by (3.7) we have that $\sum_{i=1}^n z_i(j) > \sum_{i=1}^n x_i'$. Let $T \triangleq \{i \mid n+1 \leq i \leq 2n, z_i(j) < x_i'\}$. We have from (3.7) that $\sum_{i \in I_+} z_i(j) = \sum_{i \in I_-} z_i(j)$. Using (3.23) we obtain that there exist $r \in I_+$ and $s \in I_-$ such that $z_r(j) > x_r'$ and $z_s(j) > x_s'$. Let $t$ be an arbitrary index from $T$, $\mathbf{u}_{rst} \in I_3$ such that $u_t = 2$ and

$$\lambda = \min\left(z_r(j) - x_r', z_s(j) - x_s', (x_t' - z_t(j))/2\right) .$$

Note that $\lambda > 0$. We set $\mathbf{v}(j) = \mathbf{u}_{rst}$, $\gamma(j) = \lambda$ and $\mathbf{z}(j+1) = \mathbf{z}(j) + \gamma(j)\mathbf{v}(j)$. Since $\mathbf{z}(j) \in \mathcal{F}$, it follows from the definition of $\lambda$ and $\mathbf{u}_{rst}$ that $\mathbf{z}(j+1) \in \mathcal{F}$. Since $\mathbf{z}(j)$ satisfies (3.22), we have that $x_r > x_r' \geq 0$ and $x_s > x_s' \geq 0$. Thus, $\mathbf{u}_{rst}$ is a feasible direction at $\mathbf{x}$.

Finally, it follows from the definition of $r$, $s$, $t$, $\lambda$ and $\mathbf{u}_{rst}$ that $z_i(j+1) \geq x_i' \Leftrightarrow z_i(j) \geq x_i'$ for any $1 \leq i \leq 2n$. Hence, $\mathbf{z}(j+1)$ satisfies (3.22).

**Case 3** $\sum_{i=n+1}^{2n} z_i(j) = \sum_{i=n+1}^{2n} x_i'$, $\exists n+1 \leq i \leq 2n$ such that $z_i(j) \neq x_i'$.

In this case we have that there exist $n+1 \leq s,t \leq 2n$ such that $z_s(j) < x_s'(j)$ and $z_t(j) > x_t'(j)$. Let

$$\lambda = \min\left(x_s'(j) - z_s(j), z_t(j) - x_t'(j)\right) \ .$$

Note that $\lambda > 0$. We set $\mathbf{v}(j) = \mathbf{u}_{st} \in I_1$, $\gamma(j) = \lambda$ and $\mathbf{z}(j+1) = \mathbf{z}(j) + \gamma(j)\mathbf{v}(j)$. Since $\mathbf{z}(j) \in \mathcal{F}$, it follows from the definition of $\lambda$ and $\mathbf{u}_{st}$ that $\mathbf{z}(j+1) \in \mathcal{F}$. Since $\mathbf{z}(j)$ satisfies (3.22), we have that $x_t > x_t' \geq 0$. Thus, $\mathbf{u}_{st}$ is a feasible direction at $\mathbf{x}$. Finally, it follows from the definition of $s$, $t$, $\lambda$ and $\mathbf{u}_{st}$ that $z_i(j+1) \geq x_i' \Leftrightarrow z_i(j) \geq x_i'$ for any $1 \leq i \leq 2n$. Hence, $\mathbf{z}(j+1)$ satisfies (3.22).

**Case 4** $\sum_{i=n+1}^{2n} z_i(j) = \sum_{i=n+1}^{2n} x_i'$, $\forall n+1 \leq i \leq 2n$, $z_i(j) = x_i'$, $\exists 1 \leq i \leq n$ such that $z_i(j) \neq x_i'$.

We claim that in this case there exist $1 \leq s,t \leq n$ such that $y_s = y_t$, $z_s(j) < x_s'(j)$ and $z_t(j) > x_t'(j)$. Indeed, since $\sum_{i=n+1}^{2n} z_i(j) = \sum_{i=n+1}^{2n} x_i'$, it follows from (3.7) that $\sum_{i=1}^{n} z_i(j) = \sum_{i=1}^{n} x_i'$. Combining this with (3.23) we obtain our claim. The rest of the construction of $\mathbf{z}(j+1)$ in this case is the same as in the previous case (starting from the definition of $\lambda$), with $I_1$ being replaced by $I_2$.

If none of these conditions is satisfied then $\mathbf{z}(j) = \mathbf{x}'$ and the path terminates.

By our choice of $\lambda$, at each iteration we strictly decrease the number of coordinates in $\mathbf{z}(j)$ that are different from the corresponding coordinates of $\mathbf{x}'$. Hence, the constructed path is finite. We have that $\mathbf{x}' = \mathbf{x} + \lambda'\mathbf{u} = \mathbf{x} + \sum_{i=0}^{m-1} \gamma(i)\mathbf{v}(i)$. Hence, $\mathbf{u} = \sum_{i=0}^{m-1} \frac{\gamma(i)}{\lambda'}\mathbf{v}(i)$.

**Proof of Theorem 1** The theorem follows from combining Theorem 3 and Lemma 2.

**Proof of Theorem 2** Recall that at each step in the working set selection procedure of caSMO we try the search direction $\mathbf{u}_{s'} \in I_1 \cup I_2 \cup I_3$ and this direction is chosen if it is given larger gain in the objective function than the best conjugate direction. Hence, by Lemma 2 the set of search directions $\mathcal{U}$ considered by caSMO contains a finite witness family and and the statement of the theorem follows from Theorem 3.

## 3.5 Experiments

### 3.5.1 Computation of $b$ in SVM+

Suppose that $\boldsymbol{\alpha}$ and $\boldsymbol{\beta}$ are the solution of (3.6). In this section we show how we compute the offset $b$ in the decision function (3.9).

By Karush-Kuhn-Tucker conditions,

$$\alpha_i > 0 \Rightarrow y_i\left(\mathbf{w} \cdot \mathbf{z}_i + b\right) = 1 - \left(\mathbf{w}^* \cdot \mathbf{z}_i^* + d\right). \tag{3.24}$$

Expressed in terms of the dual variables $\boldsymbol{\alpha}$ and $\boldsymbol{\beta}$, the condition (3.24) is $\alpha_i > 0 \Rightarrow y_i F_i + y_i b = 1 - \frac{1}{\gamma}f_i - d$, where $F_i = \sum_{j=1}^{n} \alpha_j y_j K_{ij}$ and $f_i = \sum_{j=1}^{n}(\alpha_j + \beta_j - C)K_{ij}^*$. The last condition is also equivalent to

$$\alpha_i > 0, y_i = 1 \Rightarrow b + d = 1 - \frac{1}{\gamma}f_i - F_i,$$

$$\alpha_i > 0, y_i = -1 \Rightarrow b - d = -1 + \frac{1}{\gamma}f_i - F_i.$$

These two conditions motivate the following way of computing $b$. Let $s_+ = \sum_{i:\alpha_i>0,y_i=1} 1 - f_i/\gamma - F_i$ and $n_+ = |\{i : \alpha_i > 0, y_i = 1\}|$, $s_- = \sum_{i:\alpha_i>0,y_i=-1} -1 + f_i/\gamma - F_i$, $n_+ = |\{i : \alpha_i > 0, y_i = -1\}|$. We set $b = (s_+/n_+ + s_-/n_-)/2$.

### 3.5.2 Description of the Datasets

We conducted our experiments on the datasets generated by `ESP` and `Tag a Tune` human computation games. `ESP` game[1] is played between two players. Both players are given the same image and they are asked to enter the tags that characterize it. The player scores points if her tags match the ones of her opponent. `Tag a Tune` game[2] is played between two players. Each player receives an audio clip of 30 seconds. The player only knows her clip and does not know the clip of her opponent. Each player sends to her opponent the tags that describe the clip. Upon receiving opponent's tags the player decides if she and her opponent are given the same clips. If both players have the right decision then both of them score points.

We used the dataset generated by `ESP` game (Von Ahn & Dabbish 2004 [307]) to perform image classification. This dataset has tagged images, with average 4.7 tags per image. We used the version (Guillaumin et al. 2009 [149]) of `ESP` dataset available at `lear.inrialpes.fr/people/guillaumin/data_iccv09.php`. In our experiments we used `DenseHue`, `DenseShift`, `HarrisHue`, and `HarrisSift` features that were supplied with the images. We generated a learning problem in the LUPI setting in the following way. Let $t_1$ and $t_2$ be two tags. The problem is to discriminate images tagged $t_1$ from the ones tagged $t_2$. We do not consider images with both $t_1$ and $t_2$ tags. The view $X$ is an image feature vector and the privileged information $X^*$ is all tags that were associated with the image, except $t_1$ and $t_2$. We removed the privileged information from the test images. The privileged information was represented as a vector with binary entries, indicating if a particular tag is associated with the image. We considered 3 binary problems with $t_1$="horse"/$t_2$="fish," $t_1$="horse"/$t_2$="bird" and $t_1$="fish"/$t_2$="bird." We denote these problems as ESP1/ESP2/ESP3, respectively. Each such problem has about 500 examples and is balanced. We used test size of 100 examples.

We used `Magnatagtune` dataset (available at `tagatune.org/Magnatagatune.html`) generated by `Tag a Tune` game (Law & Von Ahn 2009 [195]) to do music classification, This dataset has tagged audio clips, with average 4 tags per clip. Each clip is represented by measurements of its rhythm, pitch, and timbre at different intervals. We used only pitch and timbre features. The privileged information was represented as a vector with binary entries, indicating if a particular tag is associated with the clip. We generated learning problems in the LUPI setting in the same way as in `ESP` dataset. We considered 3 binary problems, "classic music" vs. "rock," "rock" vs. "electronic music" and "classic music" vs. "electronic music." We denote these problems as TAG1/TAG2/TAG3, respectively. Each problem has between 4000 and 6000 examples and is balanced. We used a test size of 2000 examples.

### 3.5.3 Comparison of `aSMO` and `caSMO`

We implemented `aSMO` and `caSMO` on the top of LIBSVM (Chang & Lin 2001 [67]). Similarly to LIBSVM, we also implemented incremental computation of gradients and shrinking. Similarly to LIBSVM, we used caching infrastructure that is provided by LIBSVM code.

In all our experiments we used radial basis function (RBF) kernel in both $X$ and $X^*$ spaces. Let $\sigma$ and $\sigma^*$ be the hyperparameters of this kernel in space $X$ and $X^*$, respectively.

---

[1] `www.gwap.com/gwap/gamesPreview/espgame/`
[2] `www.gwap.com/gwap/gamesPreview/tagatune/`

**TABLE 3.1**

Comparison of generalization error vs. stopping criterion $\tau$.

| $\tau$ | ESP1 | ESP2 | ESP3 | TAG1 | TAG2 | TAG3 |
|------|------|------|------|------|------|------|
| 1.99 | **27.16** | 26.42 | 30.67 | 1.95 | **11.76** | 7.45 |
| 1.9 | 28.42 | **26** | **30.33** | **1.91** | 12.05 | 7.49 |
| 1 | 28.75 | 27.92 | 32.17 | 1.97 | 12.27 | 7.35 |
| 0.1 | 29.42 | 29.42 | 32.25 | 2.18 | 12.34 | **7.32** |

**TABLE 3.2**

Mean running time (in seconds) of `gSMO`, `aSMO`, and `caSMO`.

| | gSMO | aSMO | caSMO |
|------|------|------|------|
| ESP1 | 12.41 | 4.21 | **0.13** |
| TAG1 | 44.39 | 4.59 | **0.1** |

Also, let $m$ and $m^*$ be a median distance between training examples in space $X$ and $X^*$, respectively. We used the $\log_{10}$-scaled range $[m-6, m+6]$ of $\sigma$, $[m^*-6, m^*+6]$ of $\sigma^*$, $[-1,5]$ of $C$ and $[-1,5]$ of $\gamma$. At each such range we took 10 equally spaced values of the corresponding parameter and generated $10 \times 10 \times 10 \times 10$ grid. We set $\kappa = 1e^{-3}$. In all experiments in this section we report the average running time, over all grid points and over 12 random draws of the training set.

`aSVM` and `caSVM` were initialized with $\alpha_i = 0$ and $\beta_i = C$ for all $1 \leq i \leq n$. With such initializations it can be verified that at the first iteration both `aSMO` and `caSMO` will choose a direction $\mathbf{u} \in I_3$ and it will hold that $\mathbf{g}^T\mathbf{u} = 2$, where $\mathbf{g}$ is the gradient of the objective function before the first iteration. Thus, the value of the stopping criterion $\tau$ should be smaller than 2. We observed (see Table 3.1) that quite a large value of $\tau$ is sufficient to obtain nearly best generalization error of SVM+. Based on this observation we ran all our experiments with $\tau = 1.9$.

Table 3.2 compares the mean running time of `aSMO`, `caSMO` with $k_{\max} = 1$, and the algorithm for solving SVM+, denoted by `gSMO`, that was used in (Vapnik & Vashist 2009 [304]). We report the results for ESP1 and TAG1 problems, but similar results were also observed for four other problems. Our results show that `aSMO` is order-of-magnitude faster than `gSMO` and `caSMO` is order-of-magnitude faster than `aSMO`. We also observed (see Figure 3.1) that the running time of `caSMO` depends on the maximal order of conjugacy $k_{\max}$. Based on our experiments, we suggest to set heuristically $k_{\max} = 3$. We used this value in our experiments

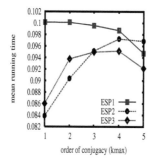
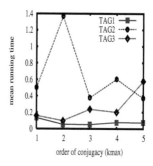

**FIGURE 3.1**

Running time (in secs) of `caSMO` as a function of the order of conjugacy.

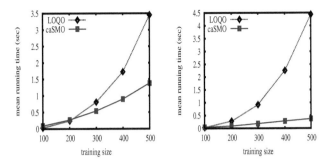

**FIGURE 3.2**
Comparison of the running time of `caSMO` and LOQO.

in Section 3.5.4. Finally, we compared (see Figure 3.2) the running time of `caSMO` with the one of LOQO off-the-shelf optimizer. We loosened the stopping criterion of LOQO so that it will output nearly the same value of objective function as `caSMO`. For very small training sets, up to 200, LOQO is competitive with `caSMO`, but for larger training sets LOQO is significantly slower.

### 3.5.4 Experiments with SVM+

We show two experiments with SVM+ and compare it with SVM and SVM*. SVM* is SVM when it is run in the privileged X* space. The experiments are for the data that is generated by `ESP` and `Tag a Tune` games. These games are described in the supplementary material. In Figures 3.3 and 3.4 we report mean test errors over 12 random draws of training and test sets. In almost all experiments SVM+ was better than SVM and relative improvement is up to 25%. In ESP1-3 and TAG3 problems SVM* is significantly better than SVM and thus the privileged space $X*$ is much better than the regular space $X$. In these problems SVM+ leverages the good privileged information to improve over SVM. In TAG1 and TAG2 problems X* space is worse than X space. Nevertheless, the results of these datasets show that SVM+ is robust to the bad privileged information. Moreover, in TAG2 dataset, despite weak X* space, SVM+ managed to extract useful information from it and outperformed SVM.

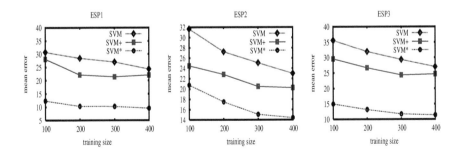

**FIGURE 3.3**
Learning curves of SVM and SVM+ in image classification.

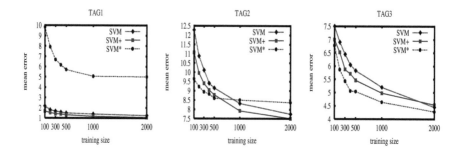

**FIGURE 3.4**
Learning curves of SVM and SVM+ in music classification.

## 3.6    Conclusions

We presented a framework of sparse conjugate direction optimization algorithms. This framework combines the ideas of the SMO algorithm and optimization using a conjugate direction. Our framework is very general and can be efficiently instantiated to any optimization problem with quadratic objective function and linear constraints. The instantiation to SVM+ gives two orders-of-magnitude running time improvement over the existing solver. We plan to instantiate the above framework to the optimization problem of SVM.

Our work leaves plenty of room for future research. It would be interesting to automatically tune the maximal order of conjugacy. Also, it is important to characterize theoretically when sparse conjugate optimization is preferable over its nonconjugate counterpart. Finally, we plan to develop optimization algorithms for SVM+ when the kernels in $X$ and/or $X^*$ space are linear.

## Acknowledgments

This work was done when the first author was with NEC Laboratories, Princeton, NJ, USA. We thank Léon Bottou for fruitful conversations.

# 4

## Conformal Predictors in Semisupervised Case

**Dmitry Adamskiy, Ilia Nouretdinov, and Alexander Gammerman**

## CONTENTS

## 4.1 Introduction

In a typical machine learning setting we are given a training set that contains description of the objects $x_1, x_2, \ldots, x_l$ and the corresponding labels $y_1, y_2, \ldots, y_l$. The aim of the machine learning algorithm is to to predict a *label* $y_{l+1}$ of a new *object* $x_{l+1}$ from the test set.

However, sometimes in addition to conventional training and test sets there are many readily available for training, but *unlabeled* examples, that can be used to improve the quality of classification. Recently these ideas have been developed under a general framework of semisupervised learning. This setting deals with learning from both labeled and unlabeled examples. The objective of this chapter is to exploit ideas of semisupervised learning and develop a corresponding conformal predictor that would allow not just to make predictions but also supply useful measures of confidence of the predictions. The basic ideas of the conformal prediction techinque were presented in (Vovk et al. 2005, 2007 [308, 140]). Conformal predictors (CP) have successfully been applied in several fields – for example, for the applications of CP in proteomics and neurosciences see (Nouretdinov et al. 2008, 2010 [246, 139, 245]).

So far, the CP technique was introduced for the supervised learning cases and here we consider possible extension to *semisupervised* problems. One of the major advantages of hedged algorithms is that they can be used for solving high-dimensional problems without requiring any parametric statistical assumptions about the source of data (unlike traditional statistical techniques); the only assumption made is the i.i.d. assumption: the examples are generated from the same probability distribution independently of each other. Actually, a weaker *exchangeability* assumption can be used to prove the property of validity: the probability distribution is invariant to the change of the order of examples.

Another advantage of the conformal predictor is that it also allows to make an estimation of confidence in the classification of individual examples.

Originally, the conformal predictor output for supervised hedged classification can be presented in one of the following ways:

- Either the user inputs a required degree (level) of certainty and the algorithm outputs a *multiprediction (prediction set)* $R_{new}$: a list of classifications that meet this confidence requirement.

- Or the algorithm outputs a so-called *forced prediction* $\hat{y}_{new}$ with its measure of *confidence*.

The second way allows interpretation of hedged prediction in a way compatible with basic prediction, but accompanies it with a confidence estimate.

Generally, high confidence in a prediction means that all alternative classifications are excluded under the i.i.d. assumption. The confidence estimates are known to be *valid* under the i.i.d. assumption, in the sense that the chance of the real classification not being on the multiprediction output list is, for example, at most 5% if the required degree of certainty is 95%.

The algorithm itself is based on testing each classification hypothesis about the new example whether it conforms to the i.i.d. assumption. This requires application of a test for randomness based on a *nonconformity measure (NCM)*, which is a way of ranking objects within a set by their relative "strangeness." The defining property of NCM is its independence of the order of examples, so any computable functions with this property can be used. Conformal predictor is valid under any choice of NCM, however, it can be more efficient (in sense of uncertainty rate, which we will discuss later) if NCM is appropriate.

The conformal predictor was originally constructed for supervised learning problems. If we now want to use unlabeled examples to improve performance in a semisupervised fashion, the following questions should be considered:

- How can the original conformal prediction be applied in a way that conserves the validity property?

- What performance measures can be used to compare the efficiency of different nonconformity measures?

- What functions are more efficient as nonconformity measures?

In this work we recall conformal prediction in the supervised case and discuss its extension to semisupervised learning.

## 4.2 Background: Conformal Prediction for Supervised Learning

### 4.2.1 Supervised Learning Problem

In the problem of *supervised learning*, we work with examples $z_i = (x_i, y_i)$. Each $x_i \in X$ is an object, normally represented as a vector, its components (dimensions) are called attributes, $y_i \in Y$ is its label. Let us assume that we are given a training set of examples

$$z_1 = (x_1, y_1), \ldots, z_l = (x_l, y_l).$$

Suppose that for a new object $x_{new}$ its real label $y_{new}$ is unknown and we are interested to get some information about it.

If $Y$ is finite, we call this problem *classification*, and if it is the set of real numbers, we call it *regression*. We will also consider an intermediate case called *complex classification*. Most examples in this chapter are related to classification unless another is stated.

### 4.2.2 Conformal Predictor

The idea of Algorithm 4 is to try every possible label $y \in Y$ as candidate for $y_{new}$ and see how well the resulting sequence of examples

$$z_1 = (x_1, y_1), \ldots, z_l = (x_l, y_l), z_{l+1} = (x_{new}, y)$$

conforms to the i.i.d. assumption (if it does, we will say it is "random"). The ideal case is where all $y \in Y$ but one leads to sequences that are not random. We can then use the remaining $y$ as a confident prediction for $x_{new}$. Checking each hypothesis about the new label whether it violates the i.i.d. assumption is done by a test for randomness based on a *nonconformity (strangeness) measure (score)* $A : (z_i, \{z_1, \ldots, z_{l+1}\}) \to \alpha_i$, which is a measure of disagreement ("nonconformity") between a finite set (subset of $X \times Y$) and its element.

The specific form of this function depends on a particular algorithm to be used and can be determined for many well-known machine learning algorithms.

---

**Algorithm 4** Conformal predictor for classification.

---

**Input:** data examples $(x_1, y_1), (x_2, y_2), \ldots, (x_l, y_l) \in X \times Y$
**Input:** a new object $x_{l+1} \in X$
**Input:** a nonconformity measure $A : (z_i, \{z_1, \ldots, z_{l+1}\}) \to \alpha_i$ on pairs $z_i \in X \times Y$
**Input(optional):** a certainty level $\gamma$
$z_1 = (x_1, y_1), \ldots, z_l = (x_l, y_l)$
**for** $y \in Y$ **do**
　$z_{l+1} = (x_{l+1}, y)$
　**for** $j$ in $1, 2, \ldots, l+1$ **do**
　　$\alpha_j = A(z_j, \{z_1, \ldots, z_l, z_{l+1}\})$
　**end for**
　$p(y) = \frac{\#\{j=1,\ldots,l+1 : \alpha_j \geq \alpha_{l+1}\}}{l+1}$
**end for**
**Output(optional):** prediction set $R_{l+1}^{\gamma} = \{y : p(y) \geq 1 - \gamma\}$
**Output:** forced prediction $\hat{y}_{l+1} = \arg\max_y \{p(y)\}$
**Output:** confidence
$conf(\hat{y}_{l+1}) = 1 - \max_{y \neq \hat{y}_{l+1}} \{p(y)\}$

---

### 4.2.3   Validity

A standard way to present the prediction results is to choose degree of certainty $\gamma < 1$ and output the ($\gamma$)-*prediction set* containing all labels with $p$-value greater than $1 - \gamma$:

$$R_{l+1}^{\gamma} = \{y : p(y) \geq 1 - \gamma\}$$

We say that this prediction is wrong if the prediction set does not include the real label $y_{new}$. The validity property from (Vovk et al. 2005 [308]) implies that probability of such error is at most $1 - \gamma$ if the i.i.d. assumption is true under the exchangeability assumption.

If the size of prediction set is 1 (or 0), we call this prediction a *certain* one, otherwise it is uncertain.

Uncertain predictions are not errors, but indications that the amount of information is not sufficient to make a unique decision at the selected level. Naturally, the higher the certainty level the more uncertain predictions will appear.

### 4.2.4   Individual Confidence

In the case of finite $Y$ (classification problem), there is an alternative way to represent the output: single ("forced") prediction and some measure of *confidence* in this prediction. It does not require the certainty level $\gamma$ as input. The single prediction is selected by largest $p$-value. We call it "forced" prediction meaning that the degree of certainty is "forced" to be low enough so that the prediction becomes certain.

The maximal degree of certainty at which the prediction is certain is called "confidence." It can be calculated as the complement to 1 of the second largest $p$-value.

For example: $Y = \{1, 2, 3\}$ and $p(1) = 0.005, p(2) = 0.05, p(3) = 0.5$, then forced prediction is 3, confidence in it is $1 - 0.05 = 0.95$. At certainty level $0.9 < 0.95$ the prediction set is $\{3\}$, so the prediction is certain at this level; at certainty level $0.99 > 0.95$ the prediction is uncertain, as the prediction set is $\{2, 3\}$, but we still say with confidence that the prediction is not 1.

In terms of confidence, validity means the following: with probability at least 0.95 either forced prediction is correct or confidence of it is not larger than 95%.

### 4.2.5   Performance Measures

Suppose that we have two or more nonconformity measures, so there are two versions of a conformal prediction algorithm and both are valid. How to check their efficiency on a real dataset, where we have a set of testing examples?

If the same certainty level is selected for several methods then we we can compare their efficiency by the *uncertainty rate*, the percentage of uncertain prediction for a fixed certainty level $\gamma$. This amount of uncertain predictions is a general characteristic of the whole prediction on a testing data set, up to the selected $\gamma$. If the selected nonconformity measure is not adequate for the problem, it will be too high. For example, if the nonconformity measure is a constant, then all predictions would be uncertain.

In terms of forced prediction with confidence, a prediction is uncertain at level $\gamma$ if its confidence is less than $\gamma$. So uncertainty rate is percentage of predictions with the individual confidence smaller than $\gamma$. To have an overall performance measure independent from $\gamma$, we can use *median (or mean) confidence*.

### 4.2.6 Nonconformity Measures Based on a Loss Function

A universal way to define a nonconformity measure is to use an underlying method of basic prediction

$$F : (X \times Y)^l \times X \to Y$$

and a loss (discrepancy) function

$$L : Y \times Y \to \mathbf{R}$$

and to use them directly as in Algorithm 5.

Although this method is usually applied for regression (see e.g., Nouretdinov et al. 2001 [247, 226]) we will see that in its pure form it is not the ideal way for classification.

---

**Algorithm 5** A scheme for NCM.

---

**Input:** data examples $z_1 = (x_1, y_1), \ldots, z_{l+1} = (x_{l+1}, y_{l+1})$
**Input:** $z_j$ (one of examples discussed in the text)
$\alpha_j = L(y_j, F(x_1, y_1, \ldots, x_{j-1}, y_{j-1}, x_{j+1}, y_{j+1}, \ldots, x_{l+1}, y_{l+1}, x_j))$

---

### 4.2.7 Example of Nonconformity Measure

The idea described above is plausible but taking only values 0 and 1 is a disadvantage of "straightforward" approach.

The threshold approach means that $\alpha_j \leq 1$ (example is strange enough) if the example is not "misclassified" by the underlying method and $\alpha_j > 1$ (example is nonstrange) if it is "misclassified." The additional gradation on each side of the threshold is made by including additional information.

For example, Algorithm 6 can be used for a 1 nearest neighbor underlying method.

---

**Algorithm 6** 1NN NCM.

---

**Input:** dataset $z_1 = (x_1, y_1), \ldots, z_{l+1} = (x_{l+1}, y_{l+1})$
**Input:** $z_j$ (one of objects above)
$d_s(j) = \min\{dist(x_j, x_k); k \neq j, y_k = y_j\}$
$d_o(j) = \min\{dist(x_j, x_k); y_k \neq y_j\}$
$\alpha_j = d_s(j)/d_o(j)$

---

This is a practically universal way to make a nonconformity measure from the underlying method: $\alpha_j$ is less than a threshold (1 in the above example) if the example $x_j$ "fits" the method and exceeds this threshold otherwise.

---

## 4.3 Conformal Prediction for Semisupervised Learning

### 4.3.1 Semisupervised Learning Problem

Semisupervised learning is a recently developed framework naturally arising in many practical tasks. Namely, for some tasks labeled examples for training could be quite expensive to obtain, whereas unlabeled ones are readily available in abundance. The classic examples of such problems are linguistic tasks where the objects are portions of texts. Unlabeled texts could be found on the Internet whereas labeling requires a skilled linguist. Therefore, the question of semisupervised learning is: could we improve the quality of the predictor given the unlabeled data?

#### 4.3.1.1 Learning on Manifolds

If there is no specific relation between the conditional distribution of a label given the object and the marginal distribution of objects then the knowledge of unlabeled data is of no help. Thus, some additional assumptions are usually made and based on the nature of these assumptions as different semisupervised algorithms arise. In what follows we present the example of utilizing two of such assumptions: manifold and cluster assumption.

*Manifold assumption*: The marginal distribution $P(x)$ is supported on a low-dimensional manifold and the conditional distribution $P(y|x)$ is smooth as a function of $x$ on the manifold.

*Cluster assumption*: We believe that the data in the same cluster is more likely to have the same label.

Usually, the distribution of unlabeled examples (and thus the manifold) is unknown to the learner. Thus, we have to model it from the (large) sample of unlabeled examples.

This is done by building the *neighborhood graph*. Note that the problem of dealing with both labeled and unlabeled data can be viewed as either labeling a partially labeled data set or as labeling the held-out test set. We shall follow the first setting as in (Belkin & Niyogi 2004 [15]), thus the neighborhood graph is built using both labeled and unlabeled points.

### 4.3.2 Conformal Predictor with External Information

The general scheme is provided by Algorithm 7. Its main difference from the conformal predictor for supervised learning (Algorithm 4) is that the nonconformity measure depends on an external argument. In particular, it may be the set of unlabeled objects.

---

**Algorithm 7** Conformal Predictor for SSL.

---

**Input:** data examples $(x_1, y_1), (x_2, y_2), \ldots, (x_l, y_l) \in X \times Y$
**Input:** a new object $x_{l+1} \in X$
**Input:** unlabeled objects $u_1, u_2, \ldots, u_s \in X$
**Input:** a nonconformity measure $A : (z_i, \{z_1, \ldots, z_{l+1}\}, U) \to \alpha_i$ on pairs $z_i \in X \times Y$ dependent on external information $U = \{u_1, u_2, \ldots, u_s\}$.
**Input(optional):** a confidence level $\gamma$
$z_1 = (x_1, y_1), \ldots, z_l = (x_l, y_l)$
**for** $y \in Y$ **do**
    $z_{l+1} = (x_{l+1}, y)$
    **for** $j$ in $1, 2, \ldots, l+1$ **do**
        $\alpha_j = A(z_j, \{z_1, \ldots, z_l, z_{l+1}\})$
    **end for**
    $p(y) = \frac{\#\{j=1,\ldots,l+1 : \alpha_j \geq \alpha_{l+1}\}}{l+1}$
**end for**
**Output(optional):** prediction set $R_{l+1}^{\gamma} = \{y : p(y) \geq 1 - \gamma\}$
**Output:** forced prediction $\hat{y}_{l+1} = \arg\max_y \{p(y)\}$
**Output:** individual confidence
$conf(\hat{y}_{l+1}) = 1 - \max_{y \neq \hat{y}_{l+1}} \{p(y)\}$

---

In the online setting it could be viewed as follows:

- The bag of unlabeled examples is presented to the learner. The learner uses this bag to fix the specific nonconformity measure from the possible family of nonconformity measures.

- Examples are presented one by one and the learner tries to predict their label.

- After the prediction is made, the label for each example is revealed.

This is very similar to the online-learning-on-graphs setting, where the graph is known to the learner and the task is to predict sequentially the labels of vertices.

Algorithm 7 preserves the validity of the original conformal predictor if the examples are randomly drawn from the bag, i.e., if all permutations are of equal probability. This is proved by the standard time-reversing argument in the same manner as for the supervised setting as it was done in (Vovk et al. 2005 [308]). Thus, even if the learner's choice of additional assumption for the semisupervised learning is not correct, the resulting predictor will still be valid.

### 4.3.3 Geo-NN: Example of NCM for Semisupervised Learning

One simple way of utilizing the unlabeled data under the manifold (and cluster) assumptions is to adjust the Nearest Neighbors NCM.

Instead of using the metric in the ambient space we can use the "geodesic" distance approximated as a distance on the neighborhood graph built using both labeled and unlabeled data. In case the manifold or cluster assumption is true it is indeed the distance we are interested in, because the conditional distribution of labels is smooth along the manifold and for a given point we want to find its nearest (labeled) neighbors on the manifold.

Obviously the construction of neighborhood graph in Algorithm 8 is done only once when the bag of unlabeled examples is presented to the learner. After this is done the distance function is fixed and we proceed with the original 1-NN conformal prediction. The

---

**Algorithm 8** Geodesic nearest neighbors NCM.

---

**Input:** dataset $z_1 = (x_1, y_1), \ldots, z_l = (x_l, y_l)$
**Input:** $z_j$ (one of objects above)
**Input:** Unlabeled data: $x_{l+1}, \ldots, x_{l+u}$
Fix the distance function $dist_0(x_i, x_j)$ in the ambient object space.
Build the neighborhood graph based on k-NN (or $\epsilon$-balls) using all the examples available (labeled and unlabeled).
**for** $i, j = 1, \ldots, l + u$ **do**
   $A_{ij} = dist_0(x_i, x_j)$ if either $x_i$ is one of the k-nearest neighbors of $x_j$ of vice versa.
   $A_{ij} = inf$ otherwise.
**end for**
Apply Floyd (or Dijkstra) algorithm to calculate distances on the graph.
Define $dist_1(x_i, x_j)$ to be the resulting distance function – length of the shortest path between $x_i$ and $x_j$ on the graph (here we make use of the "labeling partially labeled set" setting).
$d_s(j) = \min\{dist_1(x_j, x_k) | k \neq j, y_k = y_j\}$
$d_o(j) = \min\{dist_1(x_j, x_k) | y_k \neq y_j\}$
$\alpha_j = d_s(j)/d_o(j)$

---

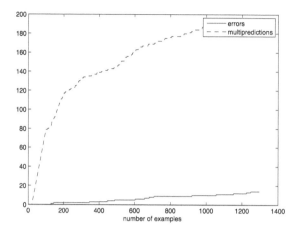

**FIGURE 4.1**
Validity of the Algorithm 8 ("geodesic"-NN NCM).

predictor is valid but we may also hope for some improvement in efficiency when the number of labeled examples is not big.

Another possible way of constructing a nonconformity measure for the semisupervised setting is to derive it from the semisupervised basic predictor in the general way mentioned above.

---

**Algorithm 9** SSL-based NCM.

**Input:** dataset $z_1 = (x_1, y_1), \ldots, z_{l+1} = (x_{l+1}, y_{l+1})$
**Input:** unlabeled objects $u_1, u_2, \ldots, u_s \in X$
**Input:** $z_j$ (one of labeled objects above)
Train a basic predictor $D(x) = D(x, z_1, \ldots, z_{j-1}, z_{j+1}, \ldots, z_{l+1}, x_j, u_1, u_2, \ldots, u_s)$
$\alpha_j = |y_j - D(x_j)|$

---

Figure 4.1 shows the number of errors and the number of multipredictions for the confidence level of 99%. We can see that the predictor is indeed valid. However, what is the performance benefit?

Figure 4.2 shows the result of applying the "geodesic-NN" conformal predictor and the 1NN-Threshold conformal predictor in the online mode to the subset of United States Postal System (USPS) dataset consisting of hand-written digits 4 and 9 (known to be rather hard to separate). We see that median confidence for the "geodesic-NN" is higher when the number of labeled examples is small and the performance levels grow as the number of examples grows. This is what one would expect from the semisupervised setting.

## 4.4   Conclusion

In this work we discussed how conformal prediction originally developed for supervised learning can be extended to a problem with other structure of input (semisupervised learning in this case).

The chapter has also shown the performance of semisupervised conformal predictors and

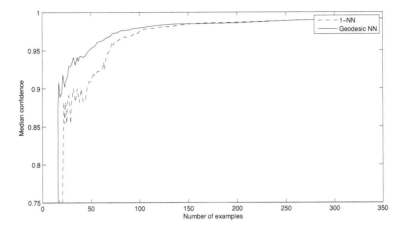

**FIGURE 4.2**
Median confidence of the Algorithm 6 (1NN threshold NCM) and Algorithm 8 ("geodesic"-NN NCM).

compared it with the results of supervised conformal prediction. The efficiency of conformal predictors can be improved by taking into account unlabeled examples when an appropriate nonconformity measure can be chosen.

Besides development of this method, a topic for future research is the extension of a conformal predictor for different types of learning such as unsupervised learning or learning with missing values. As before, the central question is related to the proof of validity and also to the assessments of various performance measures.

## Acknowledgments

We would like to thank the University Complutense de Madrid and its Programa De Visitantes Distinguidos Universidad Complutense Estancias de Professores e Investigadores Experiencia Convocatoria 2010 for their support in conducting this research. This work was also supported in part by funding from the BBSRC and ERA-NET SysBio; supported by EPSRC grant EP/F002998/1 (Practical competitive prediction); MRC grant G0802594 (Application of conformal predictors to functional magnetic resonance imaging research); VLA of DEFRA grant on Development and Application of Machine Learning Algorithms for the Analysis of Complex Veterinary Data Sets, EU FP7 grant O-PTM-Biomarkers (2008–2011); and by grant PLHRO/0506/22 (Development of New Conformal Prediction Methods with Applications in Medical Diagnosis) from the Cyprus Research Promotion Foundation.

# 5

## Some Properties of Infinite VC-Dimension Systems

**Alexey Chervonenkis**

## CONTENTS

## 5.1 Preliminaries

Let $X$ be a set, $R$ be a sigma-algebra of measurable subsets, $\mathbb{P}$ be a probability distribution defined for all elements of $R$, and $\mathbf{S}$ be some system of measurable sets $A \in X$. Let $X_l = \{x_1, x_2, \dots x_l\}$ be an independent sequence, taken from $X$ with distribution $P$. Two sets $A_1$ and $A_2$ are undifferentiable on the sequence, if and only if

$$\forall i = 1 \dots l, \quad x_i \in A_1 \iff x_i \in A_2 .$$

Now we define $\Delta^{\mathbf{S}}(X_l)$ as the number of classes of sets $A \in \mathbf{S}$, undifferentiable on $X_l$. If $\max(\log_2 \Delta^{\mathbf{S}}(X_l)) \equiv l$, we say that the system $\mathbf{S}$ has infinite VC dimension.

If $\Delta^{\mathbf{S}}(X_l)$ is measurable, then we define the entropy

$$H(\mathbf{S}, l) = \mathop{\mathbb{E}}_{X_l} \left[ \log_2 \Delta^{\mathbf{S}}(X_l) \right]$$

and the entropy per element $H(\mathbf{S}, l)/l$.

It has been shown (Vapnik & Chervonenkis 1971, 1981 [302, 303]) that, when $l$ goes to infinity, the entropy per element has the limit

$$\frac{H(\mathbf{S}, l)}{l} \stackrel{l \to \infty}{\longrightarrow} c(\mathbf{S}, P) \qquad (0 \le c \le 1)$$

and $\Delta^{\mathbf{S}}(X_l)$ also converges to $c$ in probability (due to results of this work one can see that it also converges to $c$ with probability 1).

In this work we consider the case $c > 0$.

## 5.2 Main Assertion

Let us define for a measurable set $B$ and a sequence $X_l$:

- $X_{ll}(X_l, B)$: the subsequence of the elements of $X_l$ belonging to $B$,

- $l_1(X_l, B)$: the length of this subsequence,

- $\Delta^{\mathbf{S}}(X_l, B) \overset{\text{def}}{=} \Delta^{\mathbf{S}}(X_{ll})$. When $l_1 = 0$, we define $\Delta^{\mathbf{S}}(X_l, B) = 1$.

**Theorem 4** *If* $\lim\limits_{l \to \infty} H(\mathbf{S}, l) = c > 0$, *then there exists* $T \subset X$ *such that*

$$\forall l > 0, \quad \mathbb{P}\left\{\log_2 \Delta^{\mathbf{S}}(X_l, T) = l_1(X_l, T)\right\} = 1 \quad \text{and} \quad \mathbb{P}(T) = c.$$

Note: it is evident that the probability $\mathbb{P}(T)$ cannot be more than $c$.

## 5.3    Additional Definitions

For a system $\mathbf{S}_0$ of subsets of $X$, a sequence $X_l = \{x_1, \ldots x_l\}$ of elements of $X$, and a binary sequence $Y_l = \{y_1, \ldots, y_l\}$, $y_i = 0$ or $1$, we denote

$$\mathbf{S}(\mathbf{S}_0, X_l, Y_l) = \{A \in \mathbf{S}_0 \text{ such that } \forall i, \ x_i \in A \Leftrightarrow (y_i = 1)\}.$$

In other words $\mathbf{S}(\mathbf{S}_0, X_l, Y_l)$ consists of all sets $A$ from $\mathbf{S}_0$ such that, for all $i$, $x_i$ belongs to the set $A$ if and only if $y_i = 1$. If $l = 1$, $\mathbf{S}(\mathbf{S}_0, x, 1)$ consists of all sets from $\mathbf{S}_0$ that include $x$, and $\mathbf{S}(\mathbf{S}_0, x, 0)$ consists of all sets from $\mathbf{S}_0$, which do not include $x$.

Consider

$$c(\mathbf{S}_0) = \lim_{l \to \infty} \frac{H(\mathbf{S}_0, l)}{l}$$

$$c_0(\mathbf{S}_0, x) = \lim_{l \to \infty} \frac{H(\mathbf{S}(\mathbf{S}_0, x, 0), l)}{l}$$

$$c_1(\mathbf{S}_0, x) = \lim_{l \to \infty} \frac{H(\mathbf{S}(\mathbf{S}_0, x, 1), l)}{l}$$

Since $\mathbf{S}_0 \supseteq \mathbf{S}(\mathbf{S}_0, x, 1)$, $\mathbf{S}_0 \supseteq \mathbf{S}(\mathbf{S}_0, x, 0)$, and $\mathbf{S}_0 = \mathbf{S}(\mathbf{S}_0, x, 1) \cup \mathbf{S}(\mathbf{S}_0, x, 0)$, we have

$$\Delta^{\mathbf{S}_0}(X_l) \geq \Delta^{\mathbf{S}(\mathbf{S}_0, x, 1)}(X_l)$$
$$\Delta^{\mathbf{S}_0}(X_l) \geq \Delta^{\mathbf{S}(\mathbf{S}_0, x, 0)}(X_l)$$
$$\Delta^{\mathbf{S}_0}(X_l) \leq \Delta^{\mathbf{S}(\mathbf{S}_0, x, 1)}(X_l) + \Delta^{\mathbf{S}(\mathbf{S}_0, x, 0)}(X_l)$$

It implies that

$$c(\mathbf{S}_0) = \max\{c_0(\mathbf{S}_0, x), c_1(\mathbf{S}_0, x)\}.$$

In other words, either $c = c_0$ or $c = c_1$, and, in both cases, $c$ is not less than each of $c_0$, $c_1$. However, there could be points $x$ such that.

$$c(\mathbf{S}_0) = c_0(\mathbf{S}_0, x) = c_1(\mathbf{S}_0, x). \tag{5.1}$$

We shall call such points the *points of equal branching*, and we call the set of such points the *equal branching set*

$$EB(\mathbf{S}_0) = \{x \in X \text{ such that } c(\mathbf{S}_0) = c_0(\mathbf{S}_0, x) = c_1(\mathbf{S}_0, x)\}$$

It may seem that this set is possibly empty, but we shall show that it is not empty and its measure is

$$\mathbb{P}(EB(\mathbf{S}_0)) \geqslant c(\mathbf{S}_0). \tag{5.2}$$

The proof of this fact will be given in Lemma 3.

## 5.4 The Restriction Process

We now assume (5.2) is true and construct a process to find the set $T$ defined in the main theorem. Given a sequence of independent samples $x_1, x_2, \ldots, x_i, \ldots$ distributed according to $P$, we define the process of reducing the system $\mathbf{S}$ according to the following recursive rule:

- $\mathbf{S}_0 = \mathbf{S}$, $B_0 = EB(\mathbf{S}_0)$.

- If $x_i$ does not belong to $B_{i-1}$, then

$$S_i = \begin{cases} \mathbf{S}(\mathbf{S}_{i-1}, x, 1) & \text{if } c_1(\mathbf{S}_{i-1}, x_i) = c(\mathbf{S}_{i-1}), \\ \mathbf{S}(\mathbf{S}_{i-1}, x, 0) & \text{if } c_0(\mathbf{S}_{i-1}, x_i) = c(\mathbf{S}_{i-1}). \end{cases}$$

- If $x_i$ belongs to $B_{i-1}$, then

$$S_i = \begin{cases} \mathbf{S}(\mathbf{S}_{i-1}, x, 1) & \text{with probability 0.5, or} \\ \mathbf{S}(\mathbf{S}_{i-1}, x, 0) & \text{with probability 0.5.} \end{cases}$$

- In any case, $B_i = EB(\mathbf{S}_i)$.

According to the rule, for all $i$,

$$c(\mathbf{S}_i) = c(\mathbf{S}_0) = c(\mathbf{S}),$$

and due to (5.2),

$$\mathbb{P}(EB(\mathbf{S}_0)) \geq c(\mathbf{S}).$$

It is evident that $S_i \subseteq S_{i-1}$, and it is almost evident that

$$B_i \subseteq B_{i-1}.$$

We shall show in Lemma 4 that, with probability 1,

$$\mathbb{P}(B_i) \overset{i \to \infty}{\longrightarrow} c(\mathbf{S}).$$

Now the set $T$ defined in the main theorem may be found as the intersection of all the $B_i$, $i = 1, 2, \ldots$

$$T = \bigcup_{i=1}^{\infty} B_i.$$

In general the set $T$ is not unique and the random search process described above may lead to a different result.

## 5.5 The Proof

We first prove inequality (5.2).

**Lemma 3** $\mathbb{P}(EB(\mathbf{S}_0)) \geq c(\mathbf{S}_0)$.

For a fixed sequence $X_l = \{x_1, \ldots, x_l\}$ we call a binary sequence $Y_l = \{y_1, \ldots, y_l\}$ permitted if and only if

$$\mathbf{S}(\mathbf{S}_0, X_l, Y_l) \neq \emptyset.$$

The number of different permitted $Y_l$ is just $\Delta^{\mathbf{S}_0}(X_l)$.

Now we define a binary random process of length $l$ in such a way that

1. All permitted sequences $Y_l$ are equally probable: $p(Y_l) = 1/\Delta^{\mathbf{S}_0}(X_l)$,

2. All nonpermitted sequences $Y_l$ are improbable: $p(Y_l) = 0$.

Shannon's entropy of such a process will be

$$\bar{H}(l) = -\sum_{Y_l} p(Y_l) \log_2 p(Y_l) = \log_2 \Delta^{\mathbf{S}_0}(X_l)$$

and the conditional Shannon entropy will be

$$\begin{aligned}
\bar{H}\bar{H}(y_i | y_1 \ldots y_{i-1}) = \\
-p(y_i = 0 | y_1 \ldots y_{i-1}) \log_2 p(y_i = 0 | y_1 \ldots y_{i-1}) \\
-p(y_i = 1 | y_1 \ldots y_{i-1}) \log_2 p(y_i = 1 | y_1 \ldots y_{i-1}).
\end{aligned}$$

We shall denote the conditional Shannon entropy averaged over all $y_1, \ldots, y_{i-1}$ as

$$\bar{H}(y_i | y_1 \ldots y_{i-1}) = \underset{Y_{i-1}}{\mathbb{E}} \left[ \bar{H}\bar{H}(y_i | y_1 \ldots y_{i-1}) \right].$$

The total Shannon entropy can then be presented in the form

$$\bar{H}(l) = \bar{H}(y_1) + \bar{H}(y_2 | y_1) + \cdots + \bar{H}(y_l | y_1 \ldots y_{l-1}).$$

In general the elements of the sum are not monotonic, but if we average this equation over all possible $X_l$ (assuming that $X_l$ is an independent sequence of samples each distributed according to $\mathbb{P}(x)$), they become monotonically decreasing. In fact, since Shannon's information is nonnegative, we have

$$\bar{H}(y_i | y_1 \ldots y_{i-1}) \leq \bar{H}(y_i | y_1 \ldots y_{i-2}).$$

Averaging over $X_l$,

$$\underset{X_l}{\mathbb{E}} \left[ \bar{H}(y_i | y_1 \ldots y_{i-1}) \right] \leq \underset{X_l}{\mathbb{E}} \left[ \bar{H}(y_i | y_1 \ldots y_{i-2}) \right].$$

But these averaged entropies depend only on the number of predecessors, but not on their order. Therefore,

$$\underset{X_l}{\mathbb{E}} \left[ \bar{H}(y_i | y_1 \ldots y_{i-1}) \right] \leq \underset{X_l}{\mathbb{E}} \left[ \bar{H}(y_{i-1} | y_1 \ldots y_{i-2}) \right].$$

and

$$\begin{aligned}
H(\mathbf{S}_0, l) &= \mathbb{E}[\bar{H}(l)] \\
&= \mathbb{E}[\bar{H}(y_1)] + \mathbb{E}[\bar{H}(y_2 | y_1)] + \cdots + \mathbb{E}[H(y_l | y_1 \ldots y_{l-1})] \\
&\leq l \times \mathbb{E}[\bar{H}(y_1)]
\end{aligned}$$

Then

$$\underset{X_l}{\mathbb{E}} \left[ H(y_1) \right] \geq \frac{H(\mathbf{S}_0, l)}{l}$$

and further

$$\liminf_{l\to\infty} \mathbb{E}[H(y_1)] \geq c(\mathbf{S}_0)\,. \tag{5.3}$$

If $x_1$ is not an equal branching point, then, with probability 1 (on $x_2,\dots,x_l$),

$$p(y_1 = 0) \stackrel{l\to\infty}{\Longrightarrow} 1 \quad \text{or} \quad p(y_1 = 0) \stackrel{l\to\infty}{\Longrightarrow} 0\,,$$

and thus $\bar{H}(y_1) \to 0$ and

$$\mathop{\mathbb{E}}_{x_2\dots x_l} \left[ \bar{H}(y_1) \right] \to 0$$

Then, taking into account that $\bar{H}(y_1) \leq 1$,

$$E_{X_l}[\bar{H}(y_1)] = \mathop{\mathbb{E}}_{x_1\in EB(\mathbf{S}_0)} \left[ \mathop{\mathbb{E}}_{x_2\dots x_l} [\bar{H}(y_1)] \right] + \mathop{\mathbb{E}}_{x_1\notin EB(\mathbf{S}_0)} \left[ \mathop{\mathbb{E}}_{x_2\dots x_l} [\bar{H}(y_1)] \right]\,.$$

Comparing this inequality with (5.3), we have

$$\mathbb{P}(EB(\mathbf{S}_0)) \geq c(S_0)\,.$$

The lemma is proved.

Now we return to the Restriction Process, defined above.

**Lemma 4** *With the definitions of the Restriction Process,*

$$\mathbb{P}(B_i) \stackrel{i\to\infty}{\Longrightarrow} c(\mathbf{S}) \qquad \text{with probability 1.}$$

($\mathbb{P}(B_i)$ in general is a random value, as far as the sets $B_i$ are random, being determined by preceding $x$'s and $y$'s.)

We know that $\mathbb{P}(B_i) \to c(\mathbf{S}_0)$, $\mathbb{P}(B_i) < 1$ and it monotonically decreases with $i$. This is why it is enough to show that

$$\mathbb{E}\left[\mathbb{P}(B_i)\right] \stackrel{i\to\infty}{\Longrightarrow} c(\mathbf{S})$$

Again let us consider a fixed sequence $x_1, x_2, \dots, x_l$ and declare a binary sequence $Y_l = \{y_1, \dots, y_l\}$ strongly permitted if and only if

$$c(\mathbf{S}(\mathbf{S}_0, X_l, Y_l)) = c(\mathbf{S}_0) \tag{5.4}$$

It can be easily seen that all sequences $Y_l$ possibly generated by the Restriction Process are strongly permitted, that all strongly permitted $Y_l$ sequences are permitted, but in general that not all permitted $Y_l$ sequences are strongly permitted.

Now we define a binary random process in the following recursive way:

- If $(y_1, \dots, y_{i-1}, 1)$ is strongly permitted and $(y_1, \dots, y_{i-1}, 0)$ is not, then

$$\begin{aligned} p(y_i = 1 | y_1 \dots y_{i-1}) &= 1 \quad \text{and} \\ p(y_i = 0 | y_1 \dots y_{i-1}) &= 0\,. \end{aligned}$$

- If $(y_1, \dots, y_{i-1}, 0)$ is strongly permitted and $(y_1, \dots, y_{i-1}, 1)$ is not, then

$$\begin{aligned} p(y_i = 1 | y_1 \dots y_{i-1}) &= 0 \quad \text{and} \\ p(y_i = 0 | y_1 \dots y_{i-1}) &= 1\,. \end{aligned}$$

- If both $(y_1, \ldots, y_{i-1}, 1)$ and $(y_1, \ldots, y_{i-1}, 0)$ are strongly permitted, then

$$
\begin{aligned}
p(y_i = 1 | y_1 \ldots y_{i-1}) &= 0.5 \quad \text{and} \\
p(y_i = 0 | y_1 \ldots y_{i-1}) &= 0.5 \, .
\end{aligned}
$$

Again we can calculate the Shannon entropy $\bar{H}(Y_l)$ of the binary process. But now,

$$
\bar{H}(Y_l) \leq \log_2 \Delta^{\mathbf{S}}(X_l) \, .
$$

as far as the number of different strongly permitted sequences is not greater than $\Delta^{\mathbf{S}}(X_l)$ and they are not equally probable in general.

Then

$$
\bar{\bar{H}}(y_i | y_1 \ldots y_{i-1}) = \begin{cases} 0 & \text{if } x_i \notin EB(\mathbf{S}(\mathbf{S}_0, X_{i-1}, Y_{i-1})) = B_{i-1} \\ 1 & \text{if } x_i \in EB(\mathbf{S}(\mathbf{S}_0, X_{i-1}, Y_{i-1})) = B_{i-1} \end{cases}
$$

Averaging these relations over all $X_l$ (as independent sample sequences) and using $\mathbb{P}(B_i) \geq c(\mathbf{S})$, we have

$$
H(\mathbf{S}, l) \geq \mathbb{E}[\bar{H}(Y_l)] = \mathbb{E}\Big[\sum_{i=1}^{l} \bar{\bar{H}}(y_i | y_1 \ldots y_{i-1})\Big] = \sum_{i=1}^{l} \mathbb{E}[\mathbb{P}(B_i)] \geq l\, c(\mathbf{S}) \, .
$$

Now, since $\mathbb{P}(B_i) \geq \mathbb{P}(B_{i+1})$, and $\frac{H(\mathbf{S}, l)}{l} l \xrightarrow{l \to \infty} c(\mathbf{S})$,

$$
\lim_{i \to \infty} \mathbb{E}[\mathbb{P}(B_i)] = c(\mathbf{S}) \, .
$$

The lemma is proved.

**Lemma 5** *Let $B_1 \supseteq B_2 \supseteq \cdots \supseteq B_i \ldots$ be a sequence of measurable subsets of $X$, and their intersection $Q = \bigcap_{i=1}^{\infty} B_i$ have nonzero measure $\mathbb{P}(Q) > 0$.*
*If for some $l > 0$,*

$$
\mathbb{E}\left[\log_2 \Delta^{\mathbf{S}}(X_l, B_i)\right] \xrightarrow{i \to \infty} l\, \mathbb{P}(Q)
$$

*then, for this $l$,*

$$
\mathbb{P}\left\{ \log_2 \Delta^{\mathbf{S}}(X_l, Q) = l_1(X_l, Q) \right\} = 1 \, .
$$

We can change the assumption to the equivalent one:

$$
\mathbb{P}\left\{ \log_2 \Delta^{\mathbf{S}}(X_l, B_i) = l_1(X_l, B_i) \right\} \xrightarrow{i \to \infty} 1
$$

This is true because

$$
\begin{aligned}
&l_1(X_l, B_i) \geq log \Delta^{\mathbf{S}}(X_l, B_i) \, , \\
&\mathbb{E}[l_1(X_l, B_i)] = l\, \mathbb{P}(B_i) \to l\, \mathbb{P}(Q) \, , \text{and} \\
&1 \leq \Delta^{\mathbf{S}}(X_l, B_i) \leq 2^l
\end{aligned}
$$

have a fixed finite set of possible values.

Now we define the following events:

$$
\begin{aligned}
R_1 &: \quad X_{ll}(X_l, B_i) \neq X_{ll}(X_l, Q) \\
R_2 &: \quad \log_2 \Delta^{S}(X_l, B_i) = l_1(X_l, B_i) \\
R_3 &: \quad \log_2 \Delta^{S}(X_l, Q) = l_1(X_l, Q) \\
R_4 &: \quad l_1(X_l, B_l) = l_1(X_l, Q) \text{ and } \log_2 \Delta^{S}(X_l, Q) = l_1(X_l, Q)
\end{aligned}
$$

Then

$$\mathbb{P}(R_2) \le P(R_1) + P(R_4) \le P(R_1) + P(R_3).$$
$$\mathbb{P}(R_1) \overset{i\to\infty}{\longrightarrow} 0.$$
$$\mathbb{P}(R_2) \overset{i\to\infty}{\longrightarrow} 1 \quad \text{(modified assumption)}.$$
$$\mathbb{P}(R_3) \le 1 \quad \text{does not depend on } i.$$

This is why $\mathbb{P}(R_3) = 1$.
The lemma is proved.

Now to prove the main statement it is enough to show that the sequence $B_i$ satisfies the assumption of Lemma 5 with probability 1.

**Lemma 6** *For any positive $l$, with probability 1,*

$$\underset{X_l}{\mathbb{E}} \left[\log_2 \Delta^{\mathbf{S}}(B_i, X_l)\right] \overset{i\to\infty}{\longrightarrow} l\, c(\mathbf{S})$$

(We recall that the sets $B_i$ constructed by the Restriction Process are random.)

Let us return to the construction of Lemma 4, starting from $S = S_i$.
Then

$$\underset{X_l}{\mathbb{E}} \left[\bar{H}(Y_l)\right] \ge l\, c(\mathbf{S}_i) = l\, c(\mathbf{S}).$$

Now, the number of strongly permitted sequences $y_1, \ldots, y_l$ (starting from $S_i$ and $B_i$) is not greater than $\Delta^{\mathbf{S}}(X_l, B_i)$ (this is easily shown by induction.)

So

$$l_1(X_l, B_i) \ge \log_2 \Delta^{\mathbf{S}}(X_l, B_i) \ge \bar{H}(Y_l)$$

and then

$$l\, \mathbb{P}(B_i) = \underset{X_l}{\mathbb{E}}[l_1(X_l, B_i)] \ge \underset{X_l}{\mathbb{E}}[\log_2 \Delta^{\mathbf{S}}(X_l, B_i)] \ge l\, c(\mathbf{S}).$$

According to Lemma 4, $\mathbb{P}(B_i) \to c(\mathbf{S})$ with probability 1. Therefore, the above inequality proves the lemma.

Combining the results of Lemmas 5 and 6 gives the proof of the main statement.

# Part II

# Data Science, Foundations, and Applications

# 6

## Choriogenesis: the Dynamical Genesis of Space and Its Dimensions, Controlled by Correspondence Analysis

Jean-Paul Benzécri

## CONTENTS

## 6.1 Introduction

As far as theory and practice are concerned, physics at the beginning of our 21st century is very much to be admired. However, mathematically speaking, when theory is pursued to its ultimate consequences, divergences show up relative to computation. As regards phenomena, theory is unable to respond in an unequivocal way to all of the questions raised by progress in observation of the Cosmos. Therefore, one tries to pursue simultaneously observation, experimentation, and conceiving new mathematical theories.

In the generally accepted "standard model," components predominate that are dubbed "dark" because they cannot be observed through interaction with light (the electromagnetic field), but only according to the part they are to play in the interaction of matter with the gravitational field. In this model, the fields associated with electroweak and strong interactions have become quantized, but the gravitational field has not and so, for three quarters of a century, it has been attempted to quantize it.

Our purpose is to consider the *discrete* as *fundamental relative to the continuous*, instead of considering that the discontinuous should ensue through quantization from the continuous.

Mathematically speaking, physical theories rely upon a vector space with $n$ real dimensions. From the latter one shifts for example to differential manifolds, Hilbert spaces, and later commuting relations between operators. The physical continuum has been the geometrical structure that is prior to any theory. Can we instead view the physical continuum as its outcome?

It is not the discontinuous that emerges from the continuous, from where it is deployed in quantum models. Rather, according to an intuition of Abbé Lemaître (who proposed the cosmological model bearing his name) communicated to us by (astrophysicist) J.-P. Luminet, everything should result from the pulverization of an *initial grain, coccoleiosis*.[1] Hence gravitation, the science of space, founded on the discontinuous, need not be quantized a posteriori.

One should start from graph theory, ordinal relations on a discrete finite set, and so on. In fact, correspondence analysis as applied to contingency tables (tables giving for two finite sets, $I$ and $J$, the number $k(i, j)$ of cases where $i$ is linked with $j$) has accustomed us to base or found on the discrete a continuous space ([21, 19, 196, 236, 197]). The continuous space is the space of the factorial axes, a space where the relationship between elements emerges as a proximity.

The genesis of space, *choriogenesis*,[2] relies on a random sequence of spikes: *stachyogenesis*[3] or *genesis of the spike*.

This random sequence could pass through an intermediary creation of *preorders* that do not have the structure of *spike*, the structure of spike being a particular case (as we will see) of that of a preorder. Let us therefore consider the ordinal structures upon which the genesis of space could be founded.

## 6.2　Preorder

The structure of a preordered set, or briefly a *preorder*, is to be given by: a finite set $G$, called the set of *grains*; and a finite subset $D$ of the product $G \times G$, called the set of *dipoles*. In our constructions, the hypothesis that $G$ is a finite set holds at all times.

We write: $(g, g') \in D \iff g \succeq g'$, which is read "$g$ is *anterior* or equal to $g'$."

We say that a dipole $(g, g')$ is a *strict dipole* if $g \neq g'$, which is written $g \succ g'$ (read "$g$ is strictly anterior to $g'$," one may write $g' \prec g$) and we denote $D_s$ the set of strict dipoles with $(g, g') \in D_s \iff (g, g') \in D \wedge (g \neq g')$.

We posit the existence of an initial grain, written $\lozenge[G]$, or briefly $\lozenge$, such that: $\forall g \neq \lozenge$, one has: $\lozenge \succ g$.

The word "anterior" (rather than "superior") was chosen in order to recall that there should be a single element, the initial grain, at the origin of the entire structure that we intend to build.

We define in $G$ the notion of *contiguity*: $g$ is anterior-contiguous to $g'$ if $g \succ g'$ and there is no third grain $g''$ such that: $g \succ g'' \succ g'$. We write $g \succeq g'$.

In $G$, endowed with a structure of preorder, a set $\Sigma$ of grains, indexed by an integer $n$ ranging from 0 to $\widetilde{n}$ (with $\widetilde{n} \geq 1$), will be called a *series of length $\widetilde{n}$*. We will speak of the *origin*, $\Sigma(0)$, and of the *extremity* $\Sigma(\widetilde{n})$, of the series $\Sigma$. A series will be called *descending* if each of its grains is anterior or equal to every grain of rank superior of its own.

A series of grains distinct from each other is called a *chain* if any two elements $\Sigma(n)$ and $\Sigma(n-1)$, consecutive in the series, are contiguous:

$$\forall n \, (1 \leq n \leq \widetilde{n}) \; : \; \Sigma(n) \geq \Sigma(n-1) \; \vee \; \Sigma(n-1) \geq \Sigma(n)$$

---

[1]Greek: κόκκος, kokkos, grain (of wheat); λείωσις, leiôsis, pulverization.

[2]Greek: χωρίον, khôrion, place.

[3]Greek στάχυς, stakhus: spike.

A chain will be called *descending* if each of its grains is strictly anterior to any grain of rank superior to its own.

A preorder is said to be *saturated* if the condition of transitivity is satisfied:

$$(g \succ g') \wedge (g' \succ g'') \Rightarrow g \succ g''$$

With any preorder is associated a saturated preorder, defined by *saturating* $D_s$ as follows. $D_s$ is extended to the set $D_s^+$ of dipoles defined by: there exists a descending chain $\Sigma$ with origin $g$ and extremity $g' \iff (g \succ g' \in D_s^+)$.

## 6.3 Spike

The structure of a spike is a particular case of the structure of a preorder with the supplementary condition of saturation. Therefore, a *spike* is a *finite* set $G$ of elements, called grains, with a structure of partial order written $g \succeq g'$ (read: anterior or equal to $g'$) and such that there exists a grain $\Diamond[G]$, the initial grain of G, strictly anterior to any other grain.

With a *saturated preorder* is associated a spike, defined as the quotient of $G$ by the equivalence relation:

$$g \approx g' \iff g = g' \vee (g \succ g' \wedge g' \succ g)$$

This relation is chosen such that between two distinct grains of the spike there should not be verified simultaneously two opposite order relations. In the quotient spike of a preorder $G$, the class of the initial grain in $G$ plays the part of the initial grain.

We call set of *dipoles* the subset $D$ of $G \times G$ defined by:

$$(g, g') \in D \iff g \succeq g'$$

distinguishing between $D_s$, the set of strict dipoles with $(g \neq g')$, and $D_t$, the set of trivial dipoles with $(g = g')$. For a strict dipole, we write: $g \succ g'$. For any grain $g$ other than $\Diamond[G]$, we have $\Diamond[G] \succ g$.

Given two spikes $G$ and $G'$, we call a *spike morphism* from $G$ to $G'$ a set function from $G$ into $G'$, which respects the partial order and sends the initial grain $\Diamond[G]$ onto the initial grain $\Diamond[G']$.

## 6.4 Preorder and Spike

We said that the structure of a spike is a particular case of the structure of a preorder. We have, in the structure of a preorder, the same notation that has been extended to the spike. But, for a preorder, the conditions are weaker.

Given two preorders $G$ and $G'$, we call a *preorder morphism* a set function from $G$ to $G'$, which sends a dipole onto a dipole and the initial grain onto the initial grain.

We have defined, for a preorder (therefore, in particular, for a spike) the notion of a descending series.

Any morphism from a preorder $G$ to a spike is factorized through the following morphism: we saturate the preorder by setting $g \succsim g' \iff$ there is a descending series from $g$ to $g'$.

Let $g \sim g' \iff$ simultaneously $g \gtrsim g'$ and $g' \gtrsim g$. Thus, we define an equivalence relation, because putting end to end two descending series, we have a descending series. The quotient set is given by $G'$, with $g\backslash$ for the class of $g$.

The equivalence class of of $\Diamond[G]$, denoted $\Diamond[G]\backslash$, is taken to be $\Diamond[G']$. In $G'$, the partial order relation is $g\backslash \geq g'\backslash \iff g \gtrsim g'$. Transitivity itself results from the fact that: by setting end to end two descending series one has a descending series. We will say that the spike $G'$ results from *saturating* the preorder $G$.

---

## 6.5   Geometry of the Spike

To manage the *stachyogenesis*, the *geometric* structure of the spike needs to be considered. Correspondence analysis contributes to this. But this analysis itself must be approached through a random process. Thus, in coalescing two processes, we want to end up with structures that are already considered in physics – or other new ones! Let us reemphasize that, in our constructions, the hypothesis always holds that $G$ is a *finite set*.

We have defined on $G$ the notion of *contiguity* (see p. 64). The initial grain $\Diamond[G]$ is linked to any other grain $g$ by a descending chain. There exists at least the descending series $\{\Diamond[G], g\}$ of length 1. Since $G$ is a finite set, there exists a strictly descending series of maximal length. This series is a chain since otherwise one more grain could be inserted between two consecutive noncontiguous grains.

We will call the *spindle* from $h$ to $g$ the set of the chains descending from $h$ to $g$.

*Definition of a πlate*: a *πlate* is the subset of $G$, which does not include distinct elements comparable with each other (for example $g$ and $h$ with $g \neq h$ and $g \succ h$). A subset reduced to a single element is a *πlate*; in particular, the set reduced to the initial grain is a *πlate*.

Letting $g$ be a grain distinct from $\Diamond[G]$, we call *generator* or *parent* of $g$ a grain $g'$ contiguous to $g$ and anterior to $g$. The grain $g$ is said to be a direct descendant (or a child) of the grain $g'$. The set $E$ of the parents of $g$ is a *πlate*. Indeed if two distinct elements $h$ and $k$ of $E$ were comparable with each other, with: $h \succ g$, $k \succ g$, and for example $h \succ k$, we would have $h \succ k \succ g$, and $h$ would not be contiguous to $g$. The set $E$ is called the *generating πlate* of $g$, written $\pi g[g]$. For some grains, the generating *πlate* is reduced to the initial grain.

A grain $g$ is said to be *ultimate* if it has no child (which happens to be true if and only if there does not exist a grain $g'$ to which $g$ is anterior). The set of the ultimate grains of a spike $G$ is a *πlate*, called the ultimate *πlate* of spike $G$, written $\pi u[G]$ or briefly $\pi u$. We will call the *fecundity* of a grain the number of its children.

The minimum length of the chains descending from $\Diamond[G]$ to $g$ will be called *generation depth* of $g$ (starting from $\Diamond$), written $\partial^\vee \Diamond[G](g)$, or briefly $\partial^\vee \Diamond(g)$. The symbol $\partial^\vee \Diamond$ reminds us that it is a distance measured in descending from $\Diamond$.

Among the generators $gr$ of $g$ it may happen that all do not have the same generation depth but for any of them we have $\partial^\vee \Diamond(g) \leq \partial^\vee \Diamond(gr) + 1$; the equality being true for one generator $gr$ at least.

For any spike $G$, there is defined an integer valued $\beta$ioriented distance $\partial \beta[G](g, g') = \partial \beta[G](g', g)$ as the minimum number of grains in a chain of grains going from $g$ to $g'$ (or, equivalently, from $g'$ to $g$). It suffices to verify that such a chain exists between $g$ and $g'$ and this results from the fact that any grain is connected by a chain to the initial grain.

**Remark.** The term $\beta$ioriented is used because, between two consecutive grains of the chain, the order may be "$\prec$," as well as "$\succ$;" for example $g(4) \prec g(5) \succ g(6) \prec g(7) \ldots$.

Relying upon contiguity $g \succeq g'$, one may define a correspondence, denoted $ctg[G]$, between $G$-$\pi u$ (grains not in the ultimate $\pi late$) and $G$-$\Diamond$ (grains other than the initial grain) with:

$$\forall g \in G\text{-}\pi u, \forall g' \in G\text{-}\Diamond : ctg[G](g,g') = 1 \text{ if } g \succeq g' \text{ and } 0 \text{ if not.}$$

In the table $ctg[G]$, there is no row nor column with only zeros. Therefore, correspondence analysis gives a simultaneous representation of $G$-$\pi u$ and $G$-$\Diamond$. But there may possibly be rows or columns that are identical to each other. For example, with relations $\Diamond \succeq g, \Diamond \succeq g', g \succeq g'', g' \succeq g''$, $g$ and $g'$ are not distinct in the genealogy, since both have $\Diamond$ as parent and $g''$ as child. Arising from this, one could define an equivalence relation; or assign a weight to the grains; or consider some other approach.

We will consider, in the following, an example of analysis of a $ctg[G]$ table. We have indicated that the growth of $G$ must be a random process. The random process is led itself by the random process of factor analysis (i.e., correspondence analysis). The search for a factor associated with the first eigenvalue is carried out by iterating on any initial vector, and the search for the first $p$ factors is carried out through iterating on any $p$-vector.

Processes must have a local character. Real functions on $G$ should govern local fecundity, from which the global structure results. By itself, random growth generates first a *preordered set $G$* that does not have the structure of a *spike*. We have indicated that by *saturating $G$*, and then taking the quotient $G'$ of $G$, a *spike structure* is obtained.

In the process of *choriogenesis*, the variation of the sequence of the eigenvalues issued from correspondence analysis needs to be carried through. In this sequence there will be recognition of the dimension of a space, generated by the axes associated with the first eigenvalues and, thereafter, transversal dimensions. In this way, certain models in particle theory and of fields can be joined up, such as where the points in space have a transversal structure, with strings and branes. We already indicated that it remains to find in physics, as it currently is, mathematical structures agreeing with those introduced above and then to suggest new vistas for cosmology.

We posit that we can have: for $g \neq \Diamond[G]$, the $\beta$oriented distance to the initial grain $\partial \beta[G](\Diamond[G], g)$ may be reached only by a *descending* series from $\Diamond[G]$ to $g$. This is a condition that should be satisfied in *stachyogenesis*, but how can we enforce it?

Let us comment on this postulate, considering the usual genealogies. Suppose that $g$ is a remote ancestor of $h$. On the one hand, there is a chain of ancestors descending from $g$ to $h$ and, on the other hand, there are intermediates $k$ and $k'$ such that we have:

- $g$, simultaneously anterior to $h$, to $k$ and to $k'$;

- $k$, simultaneously anterior to $h$ and to $k'$;

and we will compare the length of the given chain descending from $g$ to $h$:

- to the sum of the lengths of two chains descending from $g$ to $k$, and descending from $k$ to $h$; and

- to the sum of the lengths of three chains descending from $g$ to $k'$, descending from $k$ to $k'$, and descending from $k$ to $h$.

This last sum corresponds to a nonmonotonic path, which is descending from $g$ to $k'$, ascending from $k'$ to $k$, descending from $k$ to $h$.

If there exists a nonmonotonic chain such that the $\beta$oriented distance to the initial grain is less (or even $\leq$) than the minimum obtained by a series descending from the initial grain, we could postulate that the chain $(k', k)$ (or, more generally, any ascending subchain) be identified with a point. From this there would result a preorder; and by saturating this preorder, one will obtain a spike. The process can be continued until we obtain a spike for which the required condition is satisfied (cf. Section 6.4, Preorder and Spike, p. 65).

It should be noted that this process is necessarily finite, because the spike from which one starts is itself a finite set. But it is not true that the final $\pi late$ is respected in those successive saturations. Here is a counterexample.

Let us denote $O$ the initial grain; $G$ a grain; $T$ and $T'$ two grains that will be found on the final $\pi late$. The other grains constitute five descending chains, described as follows: $\{O, G\}$, length $L > 0$ ; $\{O, T\}$, $\{O, T'\}$, $\{G, T\}$, $\{G, T'\}$, of length $L' \leq L$; At the end of the process, there remain only two grains: $O$ the initial grain; and $G'$, another grain, which is terminal resulting from $G$, $T$, and $T'$. In addition there are also, from $O$ to $G'$, three chains: two of length $L'$ resulting from $\{O, T\}$ , $\{O, T'\}$ and one of length $L$ resulting from $\{O, G\}$.

Consider now multiple spikes. They should not be arranged in one single sequence. The resulting structure has to be able to be reached in multiple ways.

We want the choriogenesis to be ruled by a physical process, from which there should result geometric conditions, analogous to those considered below for the cogeneration cardinal; and also for the compatibility of structure between different spikes.

A pair of contiguous elements suggests, in physical terms, a dipole, with a spin. Choriogenesis could be ruled by the laws of interaction between dipoles. Even if the interaction process is a random process, there should result a regularity analogous to what we want to postulate for cogeneration, namely, compatibility between structures. This would allow correspondence analysis to extract the physical continuum.

By modifying the spike by coccogenesis, by successive random adjonctions of a grain, that is one way to proceed. In general, coccogenesis can only decrease the distance between preexisting grains (offering thereby new paths).

We will denote $\mu\nu(g)$ the *minimum cardinal of cogeneration* toward $g$, the *minimum* potentially imposed on the number of generators of a grain $g$. These bounds should be understood as fixed, according to laws conceived a priori, for the course of the coccogenesis. But these bounds can be a function of the $\beta$oriented distance to the initial grain $\delta\beta[G](\Diamond[G], g)$, of generation depth $\partial^\vee\Diamond[G](g)$ (written briefly $\partial^\vee\Diamond(g)$). The limits can be a function even, too, of the structure of the spindle descending from $\Diamond[G]$ to $g$.

Adjoining a grain modifies the structure of preorder of $G$. The most general way for *stachyogenesis* is to consider as fundamental the structure of preorder in $G$, a structure that we will then modify. Coccogenesis, adjoining a grain to a given spike, is just a particular case of this general process.

---

## 6.6 Katabasis:[4] Spikes and Filters

We will now discuss the relation between spikes and associated filter structures.

Consider multiple spikes. On the one hand, within the set of spikes considered in the genesis of the same structure there should not be more than one partial order. On the other hand, compatibility has to be characterized between the various aspects that we are developing.

According to general relativity, space-type sections form a set that is partially ordered following time. Analogously, let us define a relation between spikes, written $G_1 \subset G_2$, a spike $G_2$ is said to *surround* $G_1$ if:

• considered as a set ordered by the relation of anteriority, $G_1$ is a part of $G_2$;

---

[4]Greek: $\kappa\alpha\tau\acute{\alpha}\beta\alpha\sigma\iota\varsigma$, katabasis, descent.

- $G_1$ and $G_2$ have the same initial grain $\Diamond[G_1] = \Diamond[G_2]$;

- contiguity in $G_1$ implies contiguity in $G_2$;

- $\forall(g, g')$ with $g, g' \in G_1$ : $\partial\beta[G_2](g, g') = \partial\beta[G_1](g, g')$.

Those conditions, without the last one, only imply that, for two grains $g$ and $g'$ in $G_1$, one has: $\partial\beta[G_2](g, g') \leq \partial\beta[G_1](g, g')$, because a series of grains that are contiguous in spike $G_1$, going from $g$ to $g'$, is also a series of grains contiguous in spike $G_2$.

As regards the various spikes we will consider, it must be the case that their progression, although local, permits that when two of them, $G$ and $G'$, are given there always exists a third one $G_s$ which surrounds, simultaneously, $G$ and $G'$. There is the potential to characterize this according to the final $\pi lates$ defined above.[5]

For any grain $h$, which is not in the ultimate $\pi late$, there exists at least one series of length $\tilde{n} \geq 1$ descending from $h$ to a grain $g$ in the ultimate $\pi late$.

*Proof:* Let $h \in G$, if $h \notin \pi u[G]$, there exist series descending from $h$: a series $\{g, h\}$ of length 1 where $g$ is a child of $h$; or a series $\{g...h\}$ of length $\tilde{n}$. For any such series one has $\tilde{n} < \mathrm{Card}G$. Therefore, there exists at least one series that cannot be prolonged. But the series $\{g...h\}$ cannot be prolonged into $\{g', g...h\}$, unless $g$ has no child; that is to say, if $g \in \pi u[G]$.

For any grain $h$ in spike $G$, we will call "field of $h$ in the ultimate $\pi late$" the set $c\pi u[G](h)$ of the ultimate grains $g$ such that $g \prec h$ or $g = h$. In particular, for the initial grain $\Diamond[G]$, we have $c\pi u[G](\Diamond[G]) = \pi u[G]$. A grain $g$ is an ultimate grain, if and only if $c\pi u[G](g) = \{g\}$.

Having seen this now, consider for a given grain $h$ the set of the series descending from $h$ to a grain $g$ in the ultimate $\pi late$.

Define, for any grain $h \in G$, the value of the integer function $M\mu\gamma(h) \geq 0$ (since this function depends on $G$, one should write: $M\mu\gamma[G](h)$:

- If $h \in c\pi u[G](h)$ then $M\mu\gamma[G](h) = 0$;

- Else $M\mu\gamma[G](h) = M$aximum for $g \in c\pi u[G](h)$ of the $\mu$inimum of the length of a series descending from $h$ to $g$.

We will say that $M\mu\gamma(h)$ is the $\gamma$enealogical depth of $h$; in particular, we have $\pi u[G] = \{h \mid M\mu\gamma[G](h) = 0\}$. Since $G$ is a finite set, the function $M\mu\gamma[G](g)$ has a maximum, and we write $\int \gamma[G] = \max_{g \in G} M\mu\gamma[G](g)$.

From the viewpoint of *coccoleiosis*, we have said that $G$ must result from the pulverization of a sole *initial grain* $\Diamond[G]$ defined above. For any integer $\partial^\wedge\Diamond$, with $1 \leq \int \partial^\wedge\Diamond[G]$, we write ($\Sigma$ for sphere):

- $\Sigma(\partial^\wedge\Diamond, [G]) = \{g \mid g \in G; \partial^\wedge\Diamond[G](g) = \partial^\wedge\Diamond\}$

- $\Sigma(\partial^\wedge\Diamond, [G])$: $\Sigma$phere centered at $\Diamond[G]$ with radius $\partial^\wedge\Diamond$.

The succession of the $\Sigma$pheres suggests *Time*. The extent of a $\Sigma$phere suggests space. The variation of the set $\Sigma$ and of its cardinal, as a function of $\partial^\wedge\Diamond$, suggests what may be an expansion within the spike descending from $\Diamond[G]$.

For two grains $g'$ and $g''$ we may characterize their proximity relative to the ultimate $\pi late$ by the intersection $c\pi u[G](g') \cap c\pi u[G](g'')$. The stronger this proximity, the bigger the cardinal of that intersection. Thus, one defines a correspondence on a given $\Sigma$phere, and also between two $\Sigma$pheres.

---

[5]Following Marios Tsatsos (Tsatsos 2008 [294]) on τόποι, topoi, we must consider the spikes as objects in a category. We have defined morphisms and it remains for us to define the product.

Without reference to the ultimate $\pi late$, we may count the common threads or the grains to which $g'$ and $g''$ are both anterior, with weights that are functions of distances. Here again we have correspondence on one $\Sigma$phere, or between two $\Sigma$pheres.

As a rule, the analysis of a correspondence suggests a spatial structure.

In itself, pulverization, as any process, suggests time. The extension of the pulverization suggests space with, initially, a number of spatial dimensions, which increases with the number of the grains. The limiting case is that of a simplex with $n$ vertices, having $(n-1)$ dimensions. But if the dimensionality is not $(n-1)$, this means that the nebula of dust stretches itself out. The reciprocal proximity is not the same between any pair of grains of dust. From this there results a number of dimensions, increasing to begin with, then regressing, and stabilizing in coccoleiosis.

With growth slowing, the dimensionality (the number of dimensions) passes through a maximum, then stabilizes on the level we see. What we see, after the main dimensions (foreshadowing space), secondary, transveral dimensions should come according to the model of *anadusis*.[6]

The initial grain $\Diamond[G]$ generates grains having as sole generator the initial grain $\Diamond[G]$: the set $\mu < \Diamond[G]$. Then the grains may have as a generator either one or several grains in $\mu < \Diamond[G]$: the set $\mu \ll \Diamond[G]$; or $\Diamond[G]$ and one or several grains in $\mu < \Diamond[G]$: the set $\mu \ll \& < \Diamond[G]$.

Afterwards, we will have grains whose generators are in the sets: $\{\Diamond[G]\}$; $\mu < \Diamond[G]$; $\mu \ll \Diamond[G]$; $\mu \ll \& < \Diamond[G]$; with one generator at least which is in one of the last two sets in this list of four.

Thus, one considers that the global reality is an infinity that is approached through spikes. Two spikes being given, there must exist a third one that fulfills simultaneously the conditions of proximity satisfied by the first two. We may also say that the views we have of the global are like projections or perspectives. But while the view of a three-dimensional object is achieved by plane projections, which may all be reconciled with the structure in three dimensions, the views one has of the global are only reconciled within a number of dimensions that must be arbitrarily high. Of this infinity, which is his own work, the Creator, who is at the summit, holds the whole as if in one point, through all the rays originating from Him. For us, who are exploring the work of the Creator, what is known may only be, in proportion to what is unknown, in the relationship between the finite and the infinite, of the uncertain to the certain and, however, of the complex to the simple too.

Here, set theory offers an analogy in the theory of Filters, for which we are indebted to (mathematician) Henri Cartan. In a determined state of science, what is known cannot be more than the complement of a neighborhood of the infinite. This complement of the infinite is the complement of what is unknown to us. The compatibility of two states of science will eventually be secured by a third state that includes and rectifies both. The unknown, complementary to this third state, will be again a neighborhood of the infinite, but a neighborhood that is partially revised. It would be wise for scientists to understand progress in such a way, linking the successive states of science. These states of science are such that each is partly contradictory in itself, and cannot be definitively acquired as would be the successive floors of a tower with a solid foundation.

---

[6]Greek: ἀνάδυσις, anadusis, emergence.

## 6.7 Product of Two or More Spikes

Consider, first, two spikes $G$ and $H$. On the product $G \times H$ of the two sets of grains, we define a spike structure.

- The *initial grain*, $\Diamond[G \times H]$, is $(\Diamond[G], \Diamond[H])$.

- The order structure is defined by: $(g, h) \succeq (g', h') \iff g \succeq g'$ and $h \succeq h'$.

- Strict order is obtained on the product if and only if it holds for at least one of the coordinates:

$(g, h) \succ (g', h') \iff$
$\big((g, h) \succeq (g', h')\big) \wedge \big((g \neq g')\big) \vee (h \neq h')\big)$

- Recall that a series of distinct grains is called a chain if, for any $n$ $(1 \leq n \leq \widetilde{n})$, the two consecutive elements in the series $\Sigma(n)$ and $\Sigma(n-1)$, are contiguous.

Let us suppose that we have: a chain $\underline{Z}$ of length $\widetilde{\underline{n}}$ on $G$, and a chain $\underline{\underline{Z}}$ of length $\widetilde{\underline{\underline{n}}}$ on $H$. We will define, on $G \times H$, chains $\Sigma$ of length $(\widetilde{\underline{n}} + \widetilde{\underline{\underline{n}}})$:

- For $m$ varying from 0 to $(\widetilde{\underline{n}} + \widetilde{\underline{\underline{n}}})$, two functions $\underline{n}(m)$ and $\underline{\underline{n}}(m)$ are chosen, satisfying the following conditions:

$$\underline{n}(0) = \underline{\underline{n}}(0) = 0 \ ; \ \underline{n}(\widetilde{\underline{n}} + \widetilde{\underline{\underline{n}}}) = \widetilde{\underline{n}} \ ; \ \underline{\underline{n}}(\widetilde{\underline{n}} + \widetilde{\underline{\underline{n}}}) = \widetilde{\underline{\underline{n}}};$$

and for two consecutive values of $m$, one of these functions increases by 1 and the other does not change. From this it results that each of these two functions takes all integer values from 0 to its maximum.

- The chain $\Sigma$ is defined by:

$$\Sigma(m) = \Big(\underline{Z}(\underline{n}(m)), \underline{\underline{Z}}(\underline{\underline{n}}(m))\Big);$$

$$\Sigma(0) = (\underline{Z}(0), \underline{\underline{Z}}(0)) \, ; \, \Sigma(\widetilde{\underline{n}} + \widetilde{\underline{\underline{n}}}) = \Big(\underline{Z}(\widetilde{\underline{n}}), \underline{\underline{Z}}(\widetilde{\underline{\underline{n}}})\Big).$$

From the condition found for contiguity in $G \times H$, it results that any chain on $G \times H$ may thus be defined entangling a chain on $G$ and a chain on $H$. (The case is possible that one of the two chains may be trivially of length 0.) Besides, it results that, on the product $G \times H$, the *βioriented distance* is the total of the distances computed for each of the coordinates:

$$\partial\beta[G \times H]\big((g, h), (g', h')\big) = \partial\beta[G](g, g') + \partial\beta[H](h, h').$$

In particular, the generation depth of a grain $(g, h)$ from the initial grain: $\Diamond[G \times H] = (\Diamond[G], \Diamond[H])$ is the total of the generation depths of both coordinates:

$$\partial^\vee\Diamond[G \times H](g, h) = \partial^\vee\Diamond[G](g) + \partial^\vee\Diamond[H](h).$$

More generally, we will consider:

- the product of any number $n$ of spikes $[G_1 \times G_2 \ldots \times G_n]$;

- the initial grain: $\Diamond[G_1 \times G_2 \ldots \times G_n] = (\Diamond[G1] \ldots \Diamond[Gn])$;

- the order structure defined by $(g_1, \ldots g_n) \succeq (g'_1, \ldots g'_n) \iff g_1 \succeq g'_1 \ldots g_n \succeq g'_n$.

Contiguity results if and only if there is contiguity for one of the coordinates and equality for the others. For example, if $n = 4$: $(g_1, g_2, g_3, g_4) \succeq (g_1', g_2', g_3', g_4')$ if and only if, on the one hand, one of the following four equalities holds:

$$(g_1', g_2', g_3', g_4') = (g_1', g_2, g_3, g_4)$$
$$= (g_1, g_2', g_3, g_4)$$
$$= (g_1, g_2, g_3', g_4)$$
$$= (g_1, g_2, g_3, g_4')$$

and, on the other hand, contiguity for the distinct coordinates that means, respectively: $g_1 \succeq g_1'$, or $g_2 \succeq g_2'$, or $g_3 \succeq g_3'$, or $g_4 \succeq g_4'$.

From that, it results also that on the product $[G_1 \times G_2 \ldots \times G_n]$ the *βioriented distance* is the sum of the distances computed for each of the coordinates:

$$\partial\beta[G_1 \times G_2 \ldots \times G_n]((g_1, \ldots g_n), (g_1', \ldots g_n'))$$
$$= \partial\beta[G_1](g_1, g_1') + \ldots + \partial\beta[G_n](g_n, g_n').$$
$$\partial\beta[G_1 \times G_2 \ldots \times G_n]((g_1, \ldots g_n), (g_1', \ldots g_n')) = \partial\beta[G_1](g_1, g_1') + \ldots + \partial\beta[G_n](g_n, g_n').$$

In particular, the *generation depth* of a grain $(g_1, \ldots g_n)$ from the initial grain $\Diamond[G1 \times G_2 \ldots \times G_n] = (\Diamond[G1], \ldots \Diamond[Gn])$ is the sum of the generation depths of its $n$ coordinates:

$$\partial^\wedge\Diamond[G_1 \times G_2 \ldots \times G_n]((g_1, \ldots, g_n)) = \partial^\wedge\Diamond[G_1](g_1) + \ldots + \partial^\wedge\Diamond[G_n](g_n)$$

For the product $G \times H$, the product of the ultimate $\pi lates$ is found to be a product of spaces. If each $\pi late$ is understood as interpolating a (differentiable) manifold or variety, the product is understood as interpolating a product of varieties.

The generating $\pi late$ of $(g, h)$, that is $\pi g[(g, h)]$, is the union of $(\pi g[g], h)$ and $(g, \pi g[(h)])$. The cardinal (number of grains) is the sum of the cardinals.

## 6.8   Correspondence Analysis: Epimixia[7]

In Choriogenesis, one searches for a passage in a direction opposite to that of quantization, and without referring to a continuum, be it a differentiable variety or a vector space. We pass from a discrete set to a continuous space. We consider here a statistical method whose purpose is this passage.

We start from a correspondence table, or a function with integer values $\geq 0$: $\{k(i, j) | i \in I; j \in J\}$ on the product $I \times J$ of two finite sets $I$ and $J$. One writes: $k(i) = \sum\{k(i, j) | j \in J\}$ ; $k(j) = \sum\{k(i, j) | i \in I\}$ ; $k = \sum\{k(i, j) | i \in I; j \in J\}$. Therefore, for the frequencies: $f(i, j) = k(i, j)/k$ ; $f(i) = k(i)/k$ ; $f(j) = k(j)/k$.

In general, the analysis of a correspondence suggests a spatial structure. From a relation between two sets $I$ and $J$ one shifts to a simultaneous spatial representation of the two sets related to each other. The analysis of a symmetrical correspondence, on $I \times I$, with $\forall i, i' : k(i, i') = k(i', i)$ indicates how to consider as a spatial proximity the affinities of two elements of $I$ with each other. The correspondence analysis method was introduced for integer valued correspondences between two finite sets. We can consider, more generally, functions with positive or zero values, and positive measures (mass distributions), on the product of two spaces.

For coccoleiosis, sets of grains, spikes, $\Sigma$pheres, were all introduced, having their own ordinal structure and with ordinal relations between them. Here is a job for correspondence analyses: the process of coccoleiosis may be conducted in such a way that the analysis

---

[7]Greek: $\overset{\text{$\grave{}$}}{\epsilon}\pi\iota\mu\iota\xi\acute{\iota}\alpha$, epimixia, correspondence.

furnishes a spatial structure allowing a physical interpretation that is more precise than the physical model itself from which choriogenesis was conceived.

With the aim of following the process of coccoleiosis by correspondence analysis, we are interested in various models of tables.

In the examples considered below the correspondences are not only between finite sets but also between spaces. Our purpose is to show that the analysis of the relation introduces a new continuous representation for spaces. If one starts from finite sets, one reaches a continuum as well, so that our examples are an introduction to the finite case of coccoleiosis.

## 6.8.1 A Model of Correspondence between Finite Sets

We define a $I \times J$ table with two sets $I$ and $J$ having the same cardinal $N$. We write: $I = \{I(n)|n = 1, \ldots N\}$ ; $J = \{J(n)|n = 1, \ldots N\}$.

We consider a sequence of strictly positive masses: $\{\mu(1), \ldots \mu(n), \ldots \mu(N)\}$; $\forall n : \mu(n) > 0$ and $\sum\{\mu(n)|n = 1, \ldots N\} = 1$.

Next, we consider a correspondence table $k$ on $I \times J$:

$\forall n = 1, \ldots N : k(I(n), J(n)) = \mu(n)$;
$\forall n = 1, \ldots N, \forall m = 1, \ldots N : (n \neq m) \Rightarrow k(I(n), J(m)) = 0$;
$\forall n = 1, \ldots N : k(I(n)) = \mu(n)$.

Via the composition of the transition from $J$ towards $I$ with the transition from $I$ to $J$, there results the identical transition on $I$. Therefore, any function of mean zero on $I$ is a factor issuing from the analysis and associated with the eigenvalue 1. The same is true on $J$. The one-to-one correspondence between $I$ and $J$ associates factors on $I$ with factors on $J$. And, in the simultaneous representation of $I$ and $J$ in the space of dimension $N - 1$, the elements of the same rank are superimposed.

We may apply the general formula to compute the distributional distance between $I(n)$ and $I(n')$, taking into account that the numbers $\mu(n)$ give the average profile:

$$d^2(I(n), I(n')) =$$
$$\sum \left\{ \left( \frac{k(I(n), J(m))}{k(I(n))} - \frac{k(I(n'), J(m))}{k(I(n'))} \right)^2 / \mu(m) \mid m = 1, \ldots N \right\}.$$

In this sum, there are two nonzero terms, $m = n$, and $m = n'$. Therefore, $n \neq n' \Rightarrow d^2(I(n), I(n')) = (1/\mu(n)) + (1/\mu(n'))$.

We may thus compute the distance from an individual $I(n)$ to the origin. The origin is the mass-center of the cloud. Therefore, its profile is $\{\mu(m)\}$. For the squared distance to $I(n)$ apply the general formula:

$$\sum \left\{ \frac{(\mu(m))^2}{\mu(m)} \mid m \neq n \right\} + \frac{(\mu(n) - 1)^2}{\mu(n)}$$
$$= \sum \{\mu(m)|m \neq n\} + \frac{(\mu(n) - 1)^2}{\mu(n)} = (1/\mu(n)) - 1.$$

We may then compute the total inertia of the cloud of the individuals relative to the origin:

$$\sum \mu(n) \left( \frac{1}{\mu(n)} - 1 \right) = N - 1,$$

which agrees with the fact that the cloud, a simplex with $N$ vertices, has the same total inertia, equal to 1, in any direction.

*Example* of a spike with its initial grain $\lozenge$; $\widetilde{n}$ individuals, which are its children; and, for each of these individuals, a terminal grain of which it is the parent. Analysis of the correspondence table $ctg[G]$, between $G$-$\pi u$ (the subset of the grains that are not the ultimate

$\pi late$) and $G$-$\Diamond$ (grains other than the initial grain); thus, a table defined by the relation of contiguity. This analysis is equivalent to that of a table, following the model given above.

Here are the tables for $\tilde{n} = 5$.

$$
\begin{vmatrix}
1 & 1 & 1 & 1 & 1 & 0 & 0 & 0 & 0 & 0 \\
0 & 0 & 0 & 0 & 0 & 1 & 0 & 0 & 0 & 0 \\
0 & 0 & 0 & 0 & 0 & 0 & 1 & 0 & 0 & 0 \\
0 & 0 & 0 & 0 & 0 & 0 & 0 & 1 & 0 & 0 \\
0 & 0 & 0 & 0 & 0 & 0 & 0 & 0 & 1 & 0 \\
0 & 0 & 0 & 0 & 0 & 0 & 0 & 0 & 0 & 1
\end{vmatrix}
\quad \text{is analyzed as} \quad
\begin{vmatrix}
1/2 & 0 & 0 & 0 & 0 & 0 \\
0 & 1/10 & 0 & 0 & 0 & 0 \\
0 & 0 & 1/10 & 0 & 0 & 0 \\
0 & 0 & 0 & 1/10 & 0 & 0 \\
0 & 0 & 0 & 0 & 1/10 & 0 \\
0 & 0 & 0 & 0 & 0 & 1/10
\end{vmatrix}
$$

### 6.8.2   Models of Correspondence between Spaces

In the following examples, we look at correspondence relations between spaces. Our purpose is to show that analyzing such a relation suggests a new continuous representation for spaces. If one starts from finite sets, one reaches the continuous as well, so that our examples are an introduction to the finite case of coccoleiosis.

**Analysis of the correspondence between two circles or of a circle with itself: symmetrical correspondence**

Reminder of the reconstitution formula for finite sets $\{i|i \in I\}$ and $\{j|j \in J\}$:

$$f(i,j) = f(i)\,f(j)\left(1 + \sum_\alpha \frac{1}{\sqrt{\lambda_\alpha}} F_\alpha(i)\,G_\alpha(j)\right) \tag{6.1}$$

Denoting $\lambda_\alpha$, $F_\alpha$ and $G_\alpha$ the eigenvalue and the factors of rank $\alpha$, and denoting $\varphi_\alpha$ the factors of variance 1, one has:

$$f(i,j) = f(i)\,f(j)\left(1 + \sum_\alpha \sqrt{\lambda_\alpha}\,\varphi_\alpha(i)\,\varphi_\alpha(j)\right) \tag{6.2}$$

For a circle, one has instead of finite sets $I$ and $J$ a continuous angular coordinate ranging from 0 to $2\pi$. We will introduce $x$ and $y$ such that:

$$\int dx = \int dy = 1$$
$$f(x) = \int f(x,y)\,dy \quad ; \quad f(y) = \int f(x,y)\,dx$$
$$\int f(x,y)\,dx\,dy = \int f(x,y)\,dy = \int f(x,y)\,dx = 1$$

and for conditional probabilities:

$$f^\star(y;x) \equiv f(x,y)/f(x) \text{ with } \int f^\star(y,x)\,dy = 1$$

Using the factors $\varphi_\alpha$ of variance 1, the reconstitution formula is:

$$f(x,y) = f(x)\,f(y)\left(1 + \sum_\alpha \sqrt{\lambda_\alpha}\,\varphi_\alpha(x)\,\varphi_\alpha(y)\right) \tag{6.3}$$

One considers *symmetrical correspondences* for which the marginal densities are constants:

$$f(x,y) \equiv f(y,x) \quad ; \quad f(x) \equiv 1 \quad ; \quad f(y) \equiv 1$$

Therefore, for the conditional probabilities:

$$f^\star(y;x) = f(x,y)/f(x) = f(x,y) = f(y,x) = f^\star(x;y)$$

There is invariance by rotation: $\forall x, y, z \ : \ f(x,y) = f(x+z, y+z)$; which may be written $\forall z \ : \ f(x,y) \equiv f(x+z, y+z)$.

For conditional probability:

$$\forall x, y, z : f^\star(x; y) = f^\star(x + z; y + z); \; \forall z : f^\star(x; y) \equiv f^\star(x + z, y + z);$$
$$\forall z : \; f^\star(x; x + z) \equiv f^\star(x - z; x) \equiv f^\star(x; x - z).$$

If we define $f^{\star\Delta}(z; x) = f^\star(x + z; x) = f(x + z, s)/f(x) = f(x + z, x)$, we have $\forall x, z$ : $f^{\star\Delta}(z; x) = f^{\star\Delta}(z; x - z)$.

Taking into account the invariance by rotation, the density $f^{\star\Delta}(z; x)$ does not depend on the second variable $x$, which allows us to define $f^{\star\Delta}(\,)$:

$$\forall x, z : \; f^{\star\Delta}(z) = f^{\star\Delta}(z; x) \qquad \forall x : \; f^{\star\Delta}(z) = f^{\star\Delta}(z; x)$$

$f^{\star\Delta}(\,)$ is a function on the circle $\geq 0$ with total mass 1 and symmetric relative to 0.

We will consider through examples the case where the positive density $f^{\star\Delta}(\,)$ is a bell-shaped curve with a maximum at the origin and zero for $z = \pi$, the point diametrically opposed to the origin.

Compute the density $f(x, y)$, relying on the following equations:

$$f^{\star\Delta}(z; x) = f^\star(x + z; x) = f(x, x + z)/f(x); \text{ and here } f(x) = 1. \text{ Therefore, } f(x, x + z) = f^{\star\Delta}(z; x) = f^{\star\Delta}(z; f(x, y) = f^{\star\Delta}(y - x).$$

In the preceding expression for $f(x, y)$ the results of factor analysis will appear in conformity with the reconstitution formula 6.3.

**First example.** $f^{\star\Delta}(z) \equiv 1 + \cos(z); \; f(x, y) = 1 + \cos(y - z) = 1 + (\cos x \cos y + \sin x \sin y).$

The factors are proportional to sin and cos. It remains to find the factors and eigenvalues through computation on the variances. Functions sin and cos both have as variance $1/2$. Therefore, the factors with variance 1 are:

$$\{\varphi_1(x); \varphi_2(x)\} = \{\sqrt{2} \cos x, \sqrt{2} \sin x\}$$
$$\{\cos x, \sin x\} = \{\sqrt{1/2}\, \varphi_1(x); \sqrt{1/2}\, \varphi_2(x)\}$$

from which it results that $f(x, y) = 1 + \cos(x - y) = 1 + \cos(x) \cos(y) + \sin(x) \sin(y) = 1 + (1/2)\varphi_1(x)\varphi_1(y) + (1/2)\varphi_2(x)\varphi_2(y)$, and according to the reconstitution formula it is seen that $\sqrt{\lambda_1} = \sqrt{\lambda_2} = 1/2$, hence the two eigenvalues are equal to $1/4$ and the trace is $1/2$.

**Second example.** $f^{\star\Delta}(z) \equiv (2/3)(1 + \cos z)^2$ where the coefficient is $2/3$ in order to have 1 for the integral of $f^{\star\Delta}$ on the circle.

$$f(x, y) = (2/3)\big(1 + \cos(x - y)\big)^2 = 2/3\big(1 + 2\cos(x - y) + \cos^2(x - y)\big)$$

One applies the trigonometrical formulae $\cos 2z = 2\cos^2 z - 1$ and $\cos^2 z = (1 + \cos 2z)/2$, from which it results that:

$$f(x, y) \;= (2/3)\big(1 + 2\cos(x - y) + (1/2)(1 + \cos 2(x - y))\big)$$
$$= 1 + (4/3)\cos(x - y) + 1/3 \cos 2(x - y)$$

The factors of variance 1 are:

$$\{\varphi_1(x); \varphi_2(x); \varphi_2(x); \varphi_2(x)\} = \{\sqrt{2}\cos x; \sqrt{2}\sin x; \sqrt{2}\cos 2x; \sqrt{2}\sin 2x\}$$
$$\{\cos x; \sin x; \cos 2x; \sin 2x\} = \{\varphi_1(x)/\sqrt{2}; \varphi_2(x)/\sqrt{2}; \varphi_2(x)/\sqrt{2}; \varphi_2(x)/\sqrt{2}\}$$

from which it results that

$$f(x, y = 1(4/3)\cos(x - y) + (1/3)\cos 2(x - y)$$
$$= 1 + (4/3)\big(\cos x \cos y + \sin x \sin y\big) + (1/3)\big(\cos 2x \cos 2y + \sin 2x \sin 2y\big)$$
$$= 1 + (2/3)\big(\varphi_1(x)\varphi_1(y) + \varphi_2(x)\varphi_2(y)\big) + (1/6)\big(\varphi_3(x)\varphi_3(y) + \varphi_4(x)\varphi_4(y)\big)$$

and it is clear that $\sqrt{\lambda_1} = \sqrt{\lambda_2} = 2/3; \; \sqrt{\lambda_3} = \sqrt{\lambda_4} = 1/6$, hence $\lambda_1 = \lambda_2 = 4/9$ and $\lambda_3 = \lambda_4 = 1/36$. The trace is $17/18$.

## 6.9   Choriogenesis, Coccoleiosis, Cosmology

Correspondence Analysis allows us to find the continuous on the discrete. (See for applications, (Benzécri et al. 1982, 1986, 1992, 1993 [18, 22, 20, 23]).) But it remains to find adequate dynamics for Choriogenesis (and coccoleiosis), and to associate such dynamics with a quantum theory of fields, in which the function space is not a Hilbert space founded on a manifold, but a space founded on the discrete chorion.

Could obstacles that physics encounters, such as divergence towards the infrared (IR), divergence towards the ultraviolet (UV), dissymmetry between matter and antimatter, existence of dark matter and dark energy – could these be surmounted following the ways suggested above?

Discreteness of the spike could take care of the UV obstacle; and boundedness could allow the IR issue to be addressed. Such principles are already considered in lattice-physics. In the study of *causets*, or *causal sets*, mathematical structures are involved that are analogous to those of spikes (see, for instance, (Johnston 2010 [280])). But we emphasize the dynamical aspects of our suggestions. With a mass assigned to the grains, dark matter is accounted for; and, possibly, dissymmetry. Dark energy should be in the very process of Choriogenesis.

An analogy can be considered between Choriogenesis and generative grammars. Here, the derivation rules of the grammar should play with various probabilities. Thus, some symbols (grains) are more fecund than others. As in common language, some names attract more epithets than others, and some adjectives attract more adverbs.

Currently, experiment, observation, and computations are all involved in the "new physics," but nonetheless a physics whose mathematical laws remain analogous to those that have been handed down.

Let us consider an analogy with Coulomb's law, discovered following Newton's law. There is an analogy between both laws but Coulomb's law cannot be understood as generalizing Newton's law. Likewise, obstacles that physics encounters must be considered as presaging important new discoveries. And on the way to discoveries, it is our hope that the dreams of this chapter, involving geometry, statistics and linguistics, among other domains, will be considered indulgently.

# 7

# Geometric Data Analysis in a Social Science Research Program: The Case of Bourdieu's Sociology

**Frédéric Lebaron**

## CONTENTS

## 7.1   Introduction

There is an essential aspect of Pierre Bourdieu's work that has been somewhat neglected by those who have written about his theory and that is his constant concern for quantifying his data material and for putting his thinking in mathematical terms.[1] This chapter provides landmarks for this aspect of his work, and outlines Bourdieu's approach, at least from *La distinction* (Bourdieu 1979 [42]) onwards: namely the geometric modeling of data, based on Geometric Data Analysis (GDA).

As Bourdieu firmly expressed: "I use Correspondence Analysis very much, because I think that it is essentially a relational procedure whose philosophy fully expresses what in my view constitutes social reality. It is a procedure that 'thinks' in relations, as I try to do with the concept of field" (Foreword of the German version of *Le Métier de Sociologue*, 1991). Bourdieu's program for quantification is not an arbitrary result of historical contingencies, but the logical consequence of a set of critical experiences and reflections about the shortcomings of dominant quantitative approaches in social sciences, which led him to a conscious and systematic move towards a geometric model more adapted to his conception of the social world.

In the first part of this chapter, we rapidly describe Bourdieu's lifelong commitment to statistics (both quantification and formalization), which led him to the choice of geometric modeling of data through the use of Correspondence Analysis (CA) and Multiple Correspondence Analysis (MCA). In the second part of the chapter, we stress the role of multidimensionality in this process, with the example of *L'anatomie du goût* (Bourdieu & De Saint-Martin 1976 [49]) and *La distinction* (Bourdieu 1979 [42]). In the third part, we show that the notion of field as developed by Bourdieu is constantly made operational

---

[1]A version of the first part of this text has been published in (Robson & Sanders 2008 [272]).

through GDA, using the example of an article called *Le patronat*, or French company leaders (Bourdieu & De Saint-Martin 1978 [50]). Then, in the last section, after the examination of his last empirical work about French publishers, we try to infer from Bourdieu's practice a general sociological research program based on the use of Geometric Data Analysis.

## 7.2   Bourdieu and Statistics

As early as the "Algerian times" (the second half of the 1950s (Bourdieu 1958 [39])), Bourdieu cooperated with statisticians of the French National Institute of Statistics and Economic Studies (INSEE). He particularly did it in large-scale labor force surveys undertaken during the period of the war of independence (around 1960). Bourdieu applied his anthropological perspective to the sociological interpretation of survey data, especially the statistics of unemployment (Bourdieu, Sayad, Darbel & Seibel 1963 [53]; Garcia-Parpet 2003 [141]; Seibel 2005 [283]; Yassine, 2008 [319]).

This collaboration continued in the 1960s at the Centre de Sociologie Européenne (then directed by Raymond Aron), as reflected in the contribution to *Les héritiers* by the statistician Alain Darbel (Bourdieu & Passeron, 1964 [52]). Darbel is associated with the epoch-making calculation of the chances of access to university for the various social class categories. In *L'amour de l'art*, Bourdieu & Darbel 1966 [48] published equations of the demand for cultural goods, where cultural capital, measured according to the level of university qualification, is the main variable explaining inequalities in access to museums. But, since that time, Bourdieu was in quest of a more structural conception.

This early[2] need for a more structural conception relates to the influence of structuralism in French social sciences in the 1960s, especially with the models of linguistics and of anthropology around Claude Lévi-Strauss. For Bourdieu, it is also based on the opposition between "substantialist" and "relational" conceptions of reality, developed by the philosopher of science Ernst Cassirer. This need is also rooted, though not explicitly, in the dynamics of mathematics under the influence of Bourbaki, who was also an implicit reference-frame for specialists of the human and social sciences. Bourdieu himself often referred to the need for scientific instruments that would be capable of grasping the relational dimension of social reality. Meanwhile, the geometric approach of data analysis developed by Jean-Paul Benzécri and his school around Correspondence Analysis was emerging (see Le Roux & Rouanet 2004 [196]; Rouanet 2006 [276]). Bourdieu had been aware of it since the 1960s.

In a chapter of *Le partage des bénéfices*, a collective book written with statisticians and economists (Darras 1966 [81]), Bourdieu and Darbel evoke the limits of regression techniques in social sciences. They explicitly refer to quasi-colinearity as an important shortcoming, but they develop a more epistemological critique. Social causality amounts to the global effects of a complex structure of interrelations, which is not reducible to the combination of the "pure effects" of independent variables. As Bourdieu firmly states in *La distinction* (Bourdieu 1979 [42] p. 103): "the particular relations between a dependent variable (political opinion) and so-called independent variables such as sex, age and religion, tend to dissimulate the complete system of relations that make up the true principle of the force and form specific to the effects recorded in such and such a particular correlation."

---

[2]A first version of the famous text about *La maison kabyle* strongly influenced by Lévi-Strauss dates from 1960.

## 7.3   From Multidimensionality to Geometry

Bourdieu very soon referred to what he calls diverse species of capital: economic, cultural, social, and symbolic. His scientific objective, with this theoretical apparatus, was to counterbalance a purely economic vision of society (symbolized by microeconomists trying to expand the validity of economic man like Gary Becker). It was also, at the same time, to contest a purely idealistic vision of the cultural domain developed in cultural anthropology, structural linguistics, literary studies or philosophy, by introducing an economy of symbolic goods. He tried to integrate different dimensions of social life in the perspective of a "general economy of practices" in line with the Durkheimian project (Lebaron 2003 [199]).

At the end of the 1960s, Bourdieu turned to data analysis as being the method most in "elective affinities" with his own theory (Rouanet et al. 2000 [277]). He developed the idea that if "quantification" is to take place in sociological research, it has to be multidimensional and to aim as a first step at making operational each of the basic dimensions of social space, namely the various types of capitals, e.g., economic, cultural, social and symbolic; the next step being to combine them so as to provide a geometric model of data.

*L'anatomie du goût* (Bourdieu & De Saint-Martin 1976 [49]) is the first published application of Geometric Data Analysis methods in Bourdieu's work, republished in 1979 in *La distinction*. The data were collected through a survey done on two complementary samples, using the same basic questionnaire. The scientific objective of the work was first to provide a synthetic vision of the French social space as a global structure, and to study two subsectors more in-depth: the space of the dominant classes and the space of the middle-classes ("petite bourgeoisie"), each study being based on the analysis of an Individuals×Variables table (taken from the respective subpopulation). As can be seen in Figures 7.1 (p. 80) and 7.2 (p. 82), the main elements of the geometric modeling of data were already present in this work, as Henry Rouanet, Werner Ackermann, and Brigitte Le Roux have precisely shown in an article of the *Bulletin de méthodologie sociologique* (Rouanet et al. 2000 [277]).

CA was applied to an Individuals×Variables table, which was a common practice at the time, when the use of MCA was not yet developed. The choice of active and supplementary variables was subtle: questions on a large number of tastes and cultural practices were taken as active variables; socio-demographic and occupational information was used as supplementary elements; in the first publication, they were drawn on a transparency that could be superimposed on the first principal plane resulting from the CA. This technique of visualization gives an intuition of the sociological relations between the space of tastes (lifestyles) and the space of social positions.

The *cloud of individuals* was present in the published analysis: for specific fractions of the dominant classes the dispersion of individuals was made obvious through the contours of various subclouds (*patrons*, i.e., managers, professions libérales, i.e., liberal professions) drawn by hand. This is what will be called later "structuring factors": the cloud of individuals is systematically structured by external factors in line with the methodology of Structured Data Analysis (Le Roux & Rouanet 2004 [196]). Species of capital are "fundamental dimensions" of the space to investigate; their combination – the first principal dimensions which are interpreted – is a specific result of each analysis. The resulting global social space described in *La distinction* is three-dimensional: the first three dimensions are interpreted in terms of volume of capital, composition of capital, and seniority in the class.

When referring to the space of dominant classes or of "petite bourgeoisie" (a then bidimensional space, the "volume" dimension being controlled), the first two axes are interpreted in terms of capital composition (Axis 1) and seniority in the class (Axis 2). The analysis results in a strong sociological statement about the existence of a structural homology between the *space of lifestyles* and the *space of social positions*, both being interpreted as two aspects of the same social reality.[3] In the first principal plane of the CA of the dominant classes (Figure 7.1), one finds a first opposition between an economic pole and an intellectual one (Axis 1) and a second opposition related to age and indicators of seniority in the class.

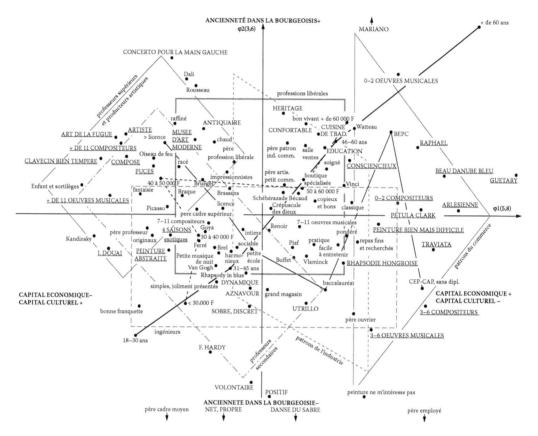

**FIGURE 7.1**
The space of the French dominant classes. Modalities and structuring factors (plane 1-2), in *La distinction* (Bourdieu 1979 [42]).

## 7.4 Investigating Fields

The "geometric modeling of data" was a practical way to combine objectification through quantitative data in a synthesis of statistical information, which is relatively close to the use of classical factor analysis, and the notion of field.

---

[3]Among today's research questions following this classical analysis we can mention here the problem of the universality of these results in other (national or historical) contexts. For scholars like Lennart Rosenlund (Rosenlund 2009 [274]), this configuration seems to be an invariant in developed capitalist societies, where the opposition between economic capital and cultural capital has become more pronounced.

As early as in the middle of the 1960s, Bourdieu formulated the concept of "field," which systematically addresses the relational aspects of social reality (Bourdieu 1966 [40]). He more completely developed his "theory of fields" in the beginning of the 1970s (Bourdieu 1971 [41]). A field is a small locus inside the global social space, which is defined by its degree of relative autonomy from other fields and the global society, by its structure, related to a specific configuration of agents, and by its dynamics. Agents in a field, even without any direct interaction, are put into objective relations, defined by the distribution of their specific resources and by a corresponding process of domination, which is distinct from the global process of social domination between classes.

The novelty of these analyses first lies in the type of data that were used in research about fields. Biographical data were collected from various biographical sources (directories, *Who's who*, etc.), in a collective process directly inspired by growing scientific practices in social history ("prosopography" from ancient and medieval history).

The second occurrence of a use of GDA by Bourdieu is a well-known article where Bourdieu and De Saint-Martin studied a population of economic elites ($n = 216$) with the help of MCA. In this article, the authors justify the central use of MCA as a way to discover and reveal a hidden relational reality that is not conscious, but nevertheless "more real" than the partial and practical perceptions of the agent (GDA as "rupture instrument" in Bachelard's tradition). They use the program MULTM from Ludovic Lebart, who is thanked for his help in a footnote. Active modalities were selected from a set of biographical data, defining the various species of capital at stake. The modalities were grouped into different headings (groups of questions), with an important number of modalities referring to social properties (from demographic characteristics to educational trajectory) and some of them to more specific assets in the economic field (positions in boards, distinctions, etc.). In particular, the following were considered:

- Demographic properties: place and date of birth, number of children, place of residence;

- Social and familial origin: profession of the father, seniority in class, presence in the Bottin Mondain (a directory of social elites);

- Educational trajectory (e.g., "grand lycée parisien");

- Professional career: (e.g., "grand corps");

- Specific positions in the field: economic power positions, membership of councils, etc.;

- Indicators of symbolic capital: official distinctions, decorations, etc.;

- Indicators of membership of engaged groups (e.g., associations).

The cloud of individuals in the first principal plane was published, with the names of economic elite members, helping the reader who "knew" some agents to have a direct intuition of the social structure of the field (Figure 7.2, p. 82). The interpreted space (Figure 7.3, p. 83) is two-dimensional (a third axis is briefly evoked). The first Axis opposes "public" to "private" positions and trajectories, the field being then dominated by technocratic managers coming from Ecole Nationale d'Administration or Polytechnique. The second Axis, again related to time, opposes "newcomers" and "established." This analysis provides a view of the structure of the field of economic elites as being defined by the relation to the State (bureaucratic capital) and by a process of dynamic competition between fractions, first defined by their seniority in the field. The new "generation" in the field is for example more often trained in private business schools, or certain engineering schools. An explanatory perspective was also present in this analysis, which aimed at understanding the space of managerial

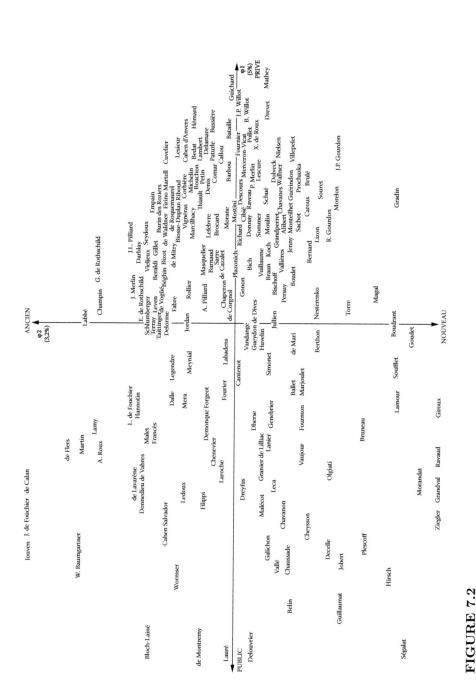

**FIGURE 7.2**
The space of the French CEOs. Individuals (plane 1-2), in *Le patronat* (Bourdieu & De Saint-Martin 1978 [50]).

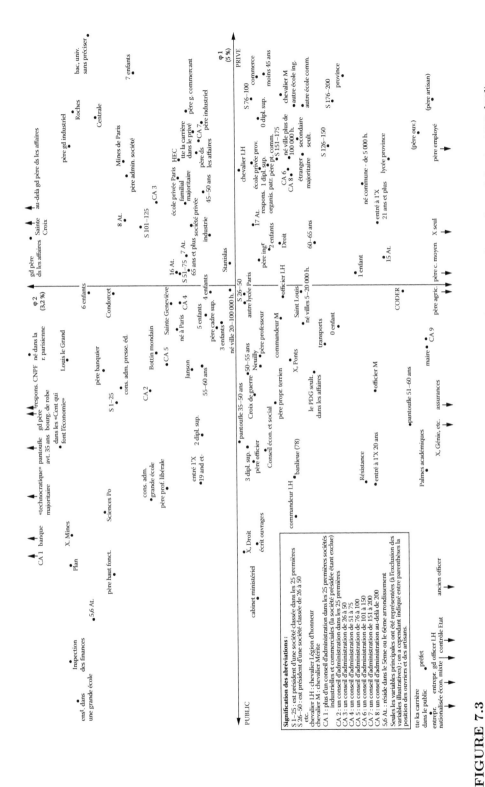

**FIGURE 7.3**

The space of the French CEOs. Cloud of modalities in plane 1-2, in *Le patronat* (Bourdieu & De Saint-Martin 1978 [50]).

strategies (for example human resources strategies) in relation to their positions in the field, and more largely "reproduction strategies" (marriage, fertility, etc.). In *La noblesse d'Etat* (Bourdieu 1989 [44]), where this article is reproduced, this analysis was combined with a study of structural homologies between the field of power in which the economic elite is included, and the field of "grandes écoles" (elite higher education institutions).

Since the late 1970s, geometric modeling has been the basis of all empirical work conducted along Bourdieu's line. It allowed Bourdieu himself to explore the major hypotheses of his theory such as: "positions [in a field] command position-takings." Two other uses of GDA techniques took place in a prosopographical study about academics in France around May 68 (Bourdieu 1984 [43]) and in an economic-sociological study about the field of private housing (Bourdieu 2000 [46]). In his last lecture at the Collège de France, in 2001, Bourdieu reiterated (Bourdieu 2001 [47] p. 70): "Those who know the principles of multiple correspondence analysis will grasp the affinities between this method of mathematical analysis and the thinking in terms of field." To be complete, one should add that Bourdieu's colleagues and followers have made intensive use of GDA methods since the middle of the 1970s. Luc Boltanski (managers, denunciation letters), Remi Lenoir (demographers, family policy), Patrick Champagne (French peasants), Monique de Saint-Martin (nobility), Annie Verger (artists), among others, have published chapters of books or articles based on MCA during the 1980s and the first half of the 1990s.

A new generation of research based on GDA in the 1990s was made visible by an article of Gisèle Sapiro about the field of French writers under German occupation published in *Actes de la recherche en sciences sociales* in 1996. Since the end of the 1980s, Swedish sociologists of education around Donald Broady and Mikael Börjesson, inspired by Bourdieu's work, were intensively using CA and MCA. Lennart Rosenlund (Rosenlund 2009 [274]) has been replicating Bourdieu's results about lifestyles in Stavanger (Norway) in the 1990s. In 1998, a conference in Cologne (Germany) paved the way to a strong new alliance between Bourdieu, sociologists referring to Bourdieu's sociological theory and statisticians interested in Bourdieu's theory like Henry Rouanet and Brigitte Le Roux. Among the outcomes of this cooperation were the analyses published in *Actes de la recherche en sciences sociales* about the field of publishers mentioned and illustrated below, and an article by (Hjellbrekke et al. 2007 [165]) putting into practice recent theoretical and technical innovations in GDA. One can also add several articles and books by (Lebaron 2001 [198]), (Duval 2004 [102]), and doctoral theses by (Denord 2003 [89]), (Börjesson 2005 [33]), (Hovden 2008 [168]) among many other applications (which could be the object of another chapter about Bourdieu's school and quantification in recent years). Recently, an article about lifestyles in the UK using specific MCA and concentration ellipses was published by Brigitte Le Roux in cooperation with a group of sociologists including Mike Savage and Alan Warde (Le Roux et al. 2008 [282]).

## 7.5 A Sociological Research Program

One can infer from Bourdieu's practice a general program based on the use of GDA in sociology. The example of his article "Une révolution conservatrice dans l'édition" (Bourdieu 1999 [45]) can be seen as the more up-to-date version of this underlying program. It is the last publication using GDA methods by Bourdieu himself. The analysis was realized in collaboration with Brigitte Le Roux and Henry Rouanet; this work followed the 1998 Cologne conference on the "Empirical Investigation of Social Space," after which cooperation between both groups of researchers began.

The geometric analysis was based on prosopographical data collected on a set of companies publishing literary books in French and translations from foreign languages ($n = 56$ individuals) (see Figures 7.4–7.7). The main method used was *specific MCA*, at that time programmed in ADDAD by Brigitte Le Roux and Jean Chiche. This technique, derived from MCA, allows to put some modalities of active questions (for example no-information or "junk" modalities) as "passive" modalities of active questions, without destroying the fundamental properties of MCA (Le Roux & Rouanet 2004 [196], (Le Roux & Rouanet 2010 [197]). As active questions ($Q = 16$), Bourdieu chose various indicators of capital (i.e., symbolic, economic and specific editorial assets, like the importance of foreign literature). As a general rule, one can state here that the sociologist has to produce the most exhaustive list of theoretically fundamental dimensions of the social resources at stake in the studied field or space.

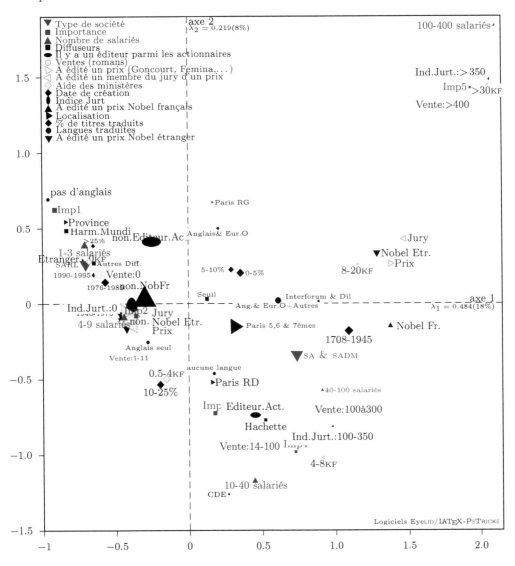

**FIGURE 7.4**

The space of French publishers. Cloud of modalities in plane 1-2 (Bourdieu 1999 [45]).

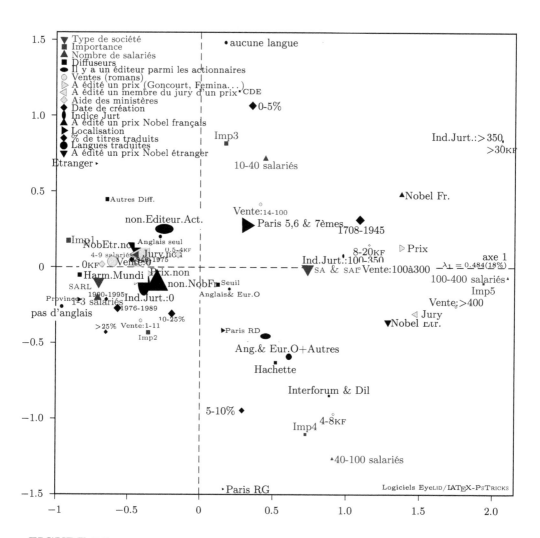

**FIGURE 7.5**
The space of French publishers. Cloud of modalities in plane 1-3 (Bourdieu 1999 [45]).

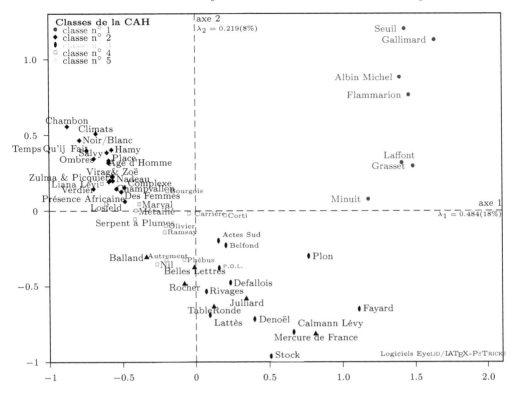

**FIGURE 7.6**
The space of French publishers. Cloud of individuals in plane 1-2 (Bourdieu 1999 [45]).

A *Euclidean clustering* was performed in order to characterize subgroups of publishers, and to "forecast" (on a qualitative basis) the future dynamics of the market (for example, the concentration processes that would reduce the number of actors, the growing domination of the economically aggressive commercial publishers).

The *sociological interpretation* insisted on the "chiasmatic" structure of the field of publishers, with a first opposition between big and small companies, and a second one between a commercial pole and a literary legitimate pole. This second Axis appears to be in homology with the classical composition Axis found in previous analyses (the classical economy versus culture Axis). The third Axis refers more to the specific importance of translations and separates two fractions of the commercial pole. Sociological interpretations assessed the relations existing between positions (related to the specific configuration of resources) and "position-takings" (editorial choices here – but it could also be political, literary or scientific choices, strategies or contents); in this article, this was mainly based on qualitative comments based on the cloud of individuals.

A research program based on this perspective would aim at:

- Showing the structure of a field or more largely of a specific social space configuration. From a statistical point of view, descriptive procedures always come first and inference only in a second phase;

- Showing structural homologies between different fields or social spaces, which has to be based on appropriate statistical and sociological interpretations of axes. Data analysis can be seen here as an interpretative practice where statistical and sociological

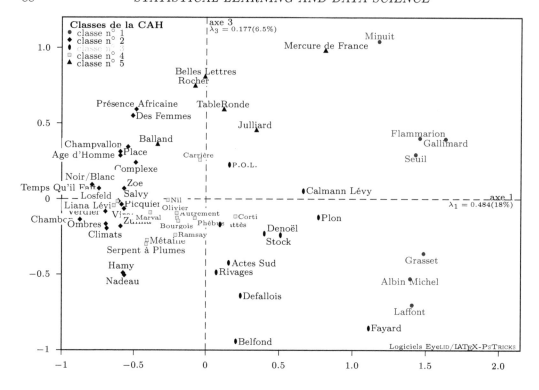

**FIGURE 7.7**
The space of French publishers. Cloud of individuals in plane 1-3 (Bourdieu 1999 [45]).

comments must converge. There is a need for more sociological meta-studies in order to cumulate converging or diverging results.

- Determining the relative autonomy of fields or social spaces, which relates to the use of comparison procedures and of "causal" sociological hypotheses, e.g., about the relations between fields or spaces;

- Studying subspaces inside a more global space. For this purpose, class specific analysis (CSA) can be an appropriate instrument (Le Roux & Rouanet 2010 [197]);

- Explaining social practices (e.g., position-taking). To quote Henry Rouanet quoting himself, Ludovic Lebart, and others, one should admit that "statistics doesn't explain anything, but gives some possible elements of explanation" (hence, statistics help in finding "candidates" for sociological explanation);

- Assessing the importance of various effects, especially field effects; this could still be very much developed, given the possibility of an integration of analysis of variance and regression techniques into the framework of GDA, such that they are more appropriate to the context of observational data;

- Studying the dynamics of a field; this could be very much developed through Euclidean clustering and/or an appropriate use of supplementary elements or, even better, structuring factors of the cloud of individuals.

## 7.6   Conclusion

Bourdieu was conscious of the shortcomings of the dominant quantitative methods in social sciences (especially regression methods), which he discovered with Alain Darbel as early as in the 1960s. He found an alternative to these methods with the geometric modeling of data, which he used for about 30 years, from the beginning of the 1970s until the late 1990s.

Bourdieu did not approve nor practice the usual rhetoric of scientific publications, presented in terms of hypotheses, empirical data, and results confirming – or failing to confirm – hypotheses. Nor did he always clearly separate the sociological interpretation and statistical one, nor did he completely formalize his theory of fields and his sociological interpretation of statistical analysis. Probably, the way his statistical practice was integrated into his sociological writing did not encourage dialog with other quantitative traditions and the clear understanding of what he did from a statistical point of view. Many researchers may find this to be regrettable.

But Bourdieu was clearly in search of a general geometric formalization; he was enthusiastic about the possibility of future integration of regression into the framework of geometric data analysis.

Bourdieu's adoption of the geometric modeling of data has therefore opened a space for a strong empirical sociological research program.

# 8

---

# Semantics from Narrative: State of the Art and Future Prospects

Fionn Murtagh, Adam Ganz, and Joe Reddington

## CONTENTS

---

## 8.1 Introduction: Analysis of Narrative

### 8.1.1 The Changing Nature of Movie and Drama

McKee (McKee 1999 [225]) bears out the great importance of the film script: "50% of what we understand comes from watching it being said." And: "A screenplay waits for the camera. ... Ninety percent of all verbal expression has no filmic equivalent."

An episode of a television series costs \$2–3 million per one hour of television, or £600k–800k for a similar series in the UK. Generally screenplays are written speculatively or commissioned, and then prototyped by the full production of a pilot episode. Increasingly, and especially availed of by the young, television series are delivered via the Internet.

Originating in one medium – cinema, television, game, online – film and drama series are increasingly migrated to another. So scriptwriting must take account of digital multimedia platforms. This has been referred to in computer networking parlance as "multiplay" and in the television media sector as a "360 degree" environment.

Cross-platform delivery motivates interactivity in drama. So-called reality TV has a considerable degree of interactivity, as well as being largely unscripted.

There is a burgeoning need for us to be in a position to model the semantics of film script, – its most revealing structures, patterns and layers. With the drive towards interactivity, we also want to leverage this work towards more general scenario analysis. Potential applications are to business strategy and planning; education and training; and science, technology, and economic development policy.

### 8.1.2   Correspondence Analysis as a Semantic Analysis Platform

For McKee (McKee 1999 [225]), film script text is the "sensory surface of a work of art" and reflects the underlying emotion or perception. Our data mining approach models and tracks these underlying aspects in the data. Our approach to textual data mining has a range of novel elements.

First, a novelty is our focus on the orientation of narrative through Correspondence Analysis ([17, 236, 277, 200]), which maps scenes (and subscenes), and words used, in a near fully automated way, into a Euclidean space representing all pairwise interrelationships. Such a space is ideal for visualization. Interrelationships between scenes are captured and displayed, as well as interrelationships between words, and mutually between scenes and words.

The starting point for analysis is frequency of occurrence data, typically the ordered scenes crossed by all words used in the script.

If the totality of interrelationships is one facet of semantics, then another is anomaly or change as modeled by a clustering hierarchy. If, therefore, a scene is quite different from immediately previous scenes, then it will be incorporated into the hierarchy at a high level. This novel view of hierarchy will be discussed further in section 8.1.3 below.

We draw on these two vantage points on semantics – viz. totality of interrelationships, and using a hierarchy to express change.

Among further work that we report on in (Murtagh et al. 2009 [238]) is the following. We devise a Monte Carlo approach to test statistical significance of the given script's patterns and structures as opposed to randomized alternatives (i.e., randomized realizations of the scenes). Alternatively we examine caesuras and breakpoints in the film script, by taking the Euclidean embedding further and inducing an ultrametric on the sequence of scenes.

### 8.1.3   Modeling Semantics via the Geometry and Topology of Information

Some underlying principles are as follows. We start with the cross-tabulation data, scenes × attributes. Scenes and attributes are embedded in a metric space. This is how we are probing the *geometry of information*, which is a term and viewpoint used by (Van Rijsbergen 2004 [298]).

We come now to a different principle: that of the *topology of information*. The particular topology used is that of hierarchy. Euclidean embedding provides a very good starting point to look at hierarchical relationships. An innovation in our work is as follows: the hierarchy takes sequence, e.g., a timeline, into account. This captures, in a more easily understood way, the notions of novelty, anomaly, or change.

Figure 8.1, left, illustrates this situation, where the anomalous or novel point is to the right. The further away the point is from the other data then the better is this approximation (Murtagh 2004 [235]). The strong triangular inequality, or ultrametric inequality, holds for tree distances: see Figure 8.1, right.

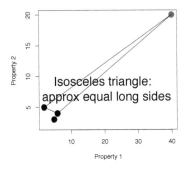 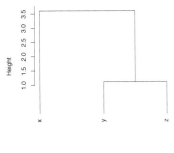

**FIGURE 8.1**

Left: The query is on the far right. While we can easily determine the closest target (among the three objects represented by the dots on the left), is the closest really that much different from the alternatives? Right: The strong triangular inequality defines an ultrametric: every triplet of points satisfies the relationship: $d(x, z) \leq \max\{d(x, y), d(y, z)\}$ for distance $d$. Cf. by reading off the hierarchy, how this is verified for all $x, y, z$: $d(x, z) = 3.5; d(x, y) = 3.5; d(y, z) = 1.0$. In addition the symmetry and positive definiteness conditions hold for any pair of points.

### 8.1.4 *Casablanca* Narrative: Illustrative Analysis

The well-known *Casablanca* movie serves as an example for us. Film scripts, such as for *Casablanca*, are partially structured texts. Each scene has metadata and the body of the scene contains dialogue and possibly other descriptive data. The *Casablanca* script was half completed when production began in 1942. The dialogue for some scenes was written while shooting was in progress. *Casablanca* was based on an unpublished 1940 screenplay (Burnet & Allison 1940 [61]). It was scripted by J.J. Epstein, P.G. Epstein, and H. Koch. The film was directed by M. Curtiz and produced by H.B. Wallis and J.L. Warner. It was shot by Warner Bros. between May and August 1942.

As an illustrative first example we use the following. A dataset was constructed from the 77 successive scenes crossed by attributes – Int[erior], Ext[erior], Day, Night, Rick, Ilsa, Renault, Strasser, Laszlo, Other (i.e., minor character), and 29 locations. Many locations were met with just once; and Rick's Café was the location of 36 scenes. In scenes based in Rick's Café we did not distinguish between "Main room," "Office," "Balcony," etc. Because of the plethora of scenes other than Rick's Café we assimilate these to just one, "other than Rick's Café," scene.

In Figure 8.2, 12 attributes are displayed; 77 scenes are displayed as dots (to avoid overcrowding of labels). Approximately 34% (for factor 1) + 15% (for factor 2) = 49% of all information, expressed as inertia explained, is displayed here. We can study interrelationships between characters, other attributes, scenes, for instance closeness of Rick's Café with Night and Int (obviously enough).

Figure 8.3 uses a sequence-constrained complete link agglomerative algorithm. It shows up scenes 9 to 10, and progressing from 39, to 40 and 41, as major changes. The sequence constrained algorithm, i.e., agglomerations are permitted between adjacent segments of scenes only, is described in (Murtagh 1985 [234]). The agglomerative criterion used, that is subject to this sequence constraint, is a complete link one.

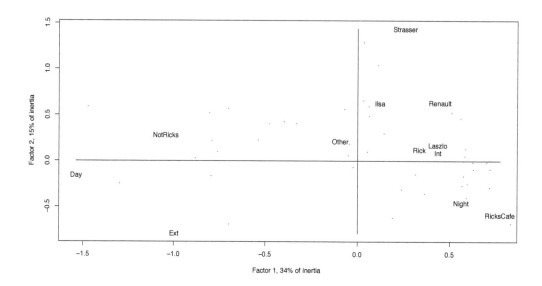

**FIGURE 8.2**
Correspondence Analysis of the *Casablanca* data derived from the script. The input data is presences/absences for 77 scenes crossed by 12 attributes. The 77 scenes are located at the dots, which are not labeled here for clarity.

### 8.1.5   Our Platform for Analysis of Semantics

Correspondence analysis supports the analysis of multivariate, mixed numerical/symbolic data. It can be used as a data analysis "platform" of choice for studying a web of interrelationships and also the evolution of relationships over time.

Correspondence Analysis is in practice *a tale of three metrics* (Murtagh 2005 [236]). The analysis is based on embedding a cloud of points from a space governed by one metric into another. Furthermore, the cloud offers vantage points of both observables and their characterizations, so – in the case of the film script – for any one of the metrics we can effortlessly pass between the space of filmscript scenes and attribute set. The three metrics are as follows.

- Chi squared, $\chi^2$, metric – appropriate for profiles of frequencies of occurrence.

- Euclidean metric, for visualization, and for static context.

- Ultrametric, for hierarchic relations and, as we use it in this work, for dynamic context.

In the analysis of semantics, we distinguish two separate aspects.

1. Context – the collection of all interrelationships.

    - The Euclidean distance makes a lot of sense when the population is homogeneous.

    - All interrelationships together provide context, relativities – and hence meaning.

2. Hierarchy tracks anomaly.

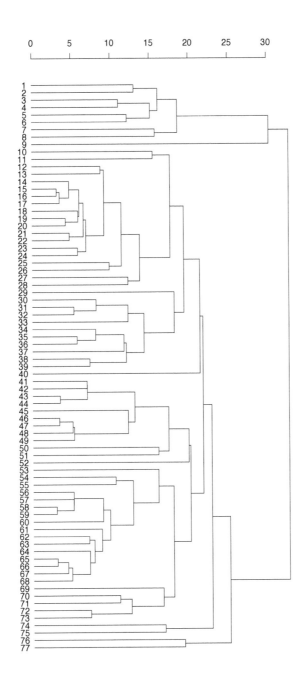

**FIGURE 8.3**

77 scenes clustered. These scenes are in sequence: a sequence-constrained agglomerative criterion is used for this. The agglomerative criterion itself is a complete link one. See (Murtagh 1985 [234]) for properties of this algorithm.

- Ultrametric distance makes a lot of sense when the observables are heterogeneous, discontinuous.

- The latter is especially useful for determining: anomalous, atypical, innovative cases.

## 8.2   Deeper Look at Semantics in the *Casablanca* Script

### 8.2.1   Text Data Mining

The *Casablanca* script has 77 successive scenes. In total there are 6710 words in these scenes. We define words as consisting of at least two letters. Punctuation is first removed. All upper case is set to lower case. We use from now on all words. We analyze frequencies of occurrence of words in scenes, so the input is a matrix crossing scenes by words.

### 8.2.2   Analysis of a Pivotal Scene, Scene 43

As a basis for a deeper look at *Casablanca* we have taken comprehensive but qualitative discussion by McKee (McKee 1999 [225]) and sought quantitative and algorithmic implementation.

*Casablanca* is based on a range of miniplots. For McKee its composition is "virtually perfect."

Following McKee (McKee 1999 [225]) we carry out an analysis of *Casablanca's* "Mid-Act Climax," Scene 43, subdivided into 11 "beats." McKee divides this scene, relating to Ilsa and Rick seeking black market exit visas, into 11 "beats."

1. Beat 1 is Rick finding Ilsa in the market.

2. Beats 2, 3, 4 are rejections of him by Ilsa.

3. Beats 5, 6 express rapprochement by both.

4. Beat 7 is guilt-tripping by each in turn.

5. Beat 8 is a jump in content: Ilsa says she will leave Casablanca soon.

6. In beat 9, Rick calls her a coward, and Ilsa calls him a fool.

7. In beat 10, Rick propositions her.

8. In beat 11, the climax, all goes to rack and ruin: Ilsa says she was married to Laszlo all along. Rick is stunned.

Figure 8.4 shows the evolution from beat to beat rather well. 210 words are used in these 11 "beats" or subscenes. Beat 8 is a dramatic development. Moving upwards on the ordinate (factor 2) indicates distance between Rick and Ilsa. Moving downwards indicates rapprochement.

In the full-dimensional space we can check some other of McKee's guidelines. Lengths of beat get shorter leading up to climax: word counts of final five beats in scene 43 are: 50 – 44 – 38 – 30 — 46. A style analysis of scene 43 based on McKee can be Monte Carlo tested against 999 uniformly randomized sets of the beats. In the great majority of cases

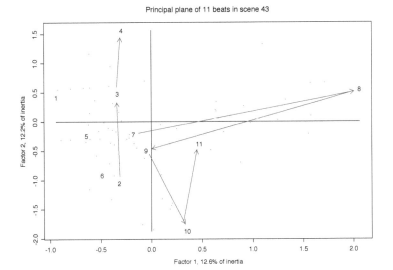

**FIGURE 8.4**
Correspondence Analysis principal plane – best Euclidean embedding in two dimensions – of scene 43. This scene is a central and indeed a pivotal one in the movie *Casablanca*. It consists of eleven subscenes, which McKee terms "beats." We discuss in the text the evolution over subscenes 2, 3 and 4; and again over subscenes 7, 8, 9, 10, and 11.

(against 83% and more of the randomized alternatives) we find the style in scene 43 to be characterized by: small variability of movement from one beat to the next; greater tempo of beats; and high mean rhythm.

The planar representation in Figure 8.4 accounts for approximately 12.6% + 12.2% = 24.8% of the inertia, and hence the total information. We will look at the evolution of this scene, scene 43, using hierarchical clustering of the full-dimensional data – but based on the relative orientations, or correlations with factors. This is because of what we have found in Figure 8.4, viz., *change* of direction is most important.

Figure 8.5 shows the hierarchical clustering, based on the sequence of beats. Input data are of full dimensionality so there is no approximation involved. Note the caesura in moving from beat 7 to 8, and back to 9. There is less of a caesura in moving from 4 to 5 but it is still quite pronounced.

Further discussion of these results can be found in (Murtagh et al. 2009 [238]).

## 8.3 From Filmscripts to Scholarly Research Articles

### 8.3.1 Introduction

Our objectives in this work include the following:

- Assess how well a part of an article stands in for the article itself. We use abstracts and also the reference sections.

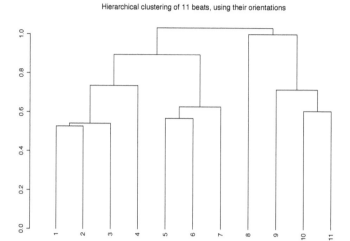

Hierarchical clustering of 11 beats, using their orientations

**FIGURE 8.5**
Hierarchical clustering of sequence of beats in scene 43 of *Casablanca*. Again, a sequence-constrained complete link agglomerative clustering algorithm is used. The input data is based on the full dimensionality Euclidean embedding provided by the Correspondence Analysis. The relative orientations (defined by correlations with the factors) are used as input data.

- Evaluate how salient sentences and/or paragraphs can be determined. Evaluate how well sentences or paragraphs can represent the article or sections of the article.

- Determine throughlines or "backbones" of the narrative in an article.

- Map out the narrative within an article.

Our medium term research program includes the following. We wish to map out the narrative between articles, or the "geneology of ideas" in this set of articles. Furthermore, we wish to specify the best attributes to characterize a large set of articles with a view towards inferring a mapping onto a set of performance "metrics."

### 8.3.2    Selection of Data

We selected a number of articles from the one theme area in order to study the structure of narrative and to seek particularly interesting semantic elements. The articles selected deal with neuro-imaging studies of visual awareness or other cognitive alternatives in early blind humans, all from the journal, NeuroImage. Some authors are shared between articles, and other authorships are not.

The selected articles were as follows: ([299, 83, 7, 26, 262]).

Minor headings were used with the subsequent paragraph and major headings were not used at all. See the caption of Table 8.1 for further details.

**TABLE 8.1**

Paragraphs included preceding subsection titles but not main titles. One paragraph was used for the Acknowledgments. One paragraph was also used for all citations. The article title was not used. The abstract in not included above: if it were the numbers of sections in each article would be increased by one. Table captions were not used, given that they were very short. Nor were tables themselves used. However figure captions, being extensive, were used, with generally each caption being taken as a paragraph in the sequence in which they are first referenced. (See text for further details here.)

| Reference | No. sections | No. paragraphs | Total words | Unique words |
|---|---|---|---|---|
| (Vanlierde et al. 2003 [299]) | 7 | 51 | 8067 | 1534 |
| (De Volder et al. 2001 [83]) | 6 | 38 | 6776 | 1408 |
| (Arno et al. 2001 [7]) | 6 | 60 | 8247 | 1534 |
| (Bittar et al. 1999 [26]) | 6 | 23 | 3891 | 999 |
| (Ptito et al. 1999 [262]) | 7 | 24 | 5167 | 1255 |

### 8.3.3 Study of All Five Articles

Using Vanlierde et al. (Vanlierde et al. 2003 [299]), De Volder et al. (De Volder et al. 2001 [83]), Arno et al. (Arno et al. 2001 [7]), Bittar et al. (Bittar et al. 1999 [26]) and Ptito et al. (Ptito et al. 1999 [262]), we took the sections used previously, for each article. We left Acknowledgments and, in a first exploration References, aside. In a second study below we use the References.

Figures were treated as follows. For Vanlierde et al. (Vanlierde et al. 2003 [299]) the captions of the 3 figures were each taken as in situ paragraphs, in the context of the relevant section. The same was done for the 3 figures in De Volder et al. (De Volder et al. 2001 [83]). For Arno et al. (Arno et al. 2001 [7]), 2 figures were used and the third was not due to the caption for the latter being very brief. For Bittar et al. (Bittar et al. 1999 [26]), the 3 figure captions were relatively brief and were in succession, so all were combined into one in situ paragraph. For Ptito et al. (Ptito et al. 1999 [262]) all 5 short figure captions were combined into one paragraph.

We note that there is overlap between authors in the first three of the selected articles, Vanlierde et al. (Vanlierde et al. 2003 [299]), De Volder et al. (De Volder et al. 2001 [83]), and Arno et al. (Arno et al. 2001 [7]); and a different cluster with mutual overlap in the case of the final two of the selected articles, Bittar et al. (Bittar et al. 1999 [26]) and Ptito et al. (Ptito et al. 1999 [262]). Thus, a first question is to see if we can recover the distinction between the two clusters of authors. This we will do on the basis of the terminology used, notwithstanding the relatively close proximity of theme in all cases.

A second issue we will address is to know how close sections stand relative to the overall article. Is the article's terminology very similar globally to the terminology of the individual sections? If not, one could argue that the coherence of the article leaves something to be desired. Or one could argue that there is something anomalous about the section. To avoid an uninteresting case of the latter, we decided to leave the Acknowledgment sections completely aside.

A third issue we will address is the following. All articles selected are well endowed with their references. How many references are cited is of course strongly associated with the subdiscipline, in this case the particular life sciences one. So what we look for is the relationship between references sections of the selected articles relative to the main content of the articles. Do we find the main distinctions between the two author clusters, for instance?

Could we adopt the same approach for the abstracts? This would lead to some provisional

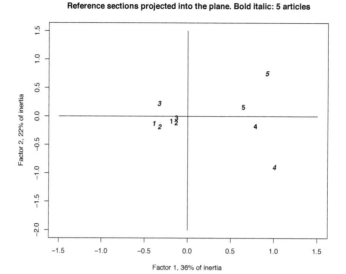

Reference sections projected into the plane. Bold italic: 5 articles

**FIGURE 8.6**

The Correspondence Analysis principal factor plane with the references sections of all 5 articles projected onto the plane. For the analysis they are "passive" in that this projection takes place after the factors have been determined.

conclusions for this particular set of articles in regard to references alone, and/or abstracts, providing proxies for the main article content.

In this chapter we will address the latter issues.

### 8.3.4    Reference Sections as Proxies for Document Content

The numbers of nonunique words in the reference sections of the 5 articles are: 1960, 1589, 1563, 1044, and 1669.

Figure 8.6 shows that the reference section tracks quite well the main body of the articles. Because they do differ in having lots more, and different, terms – notably co-author names, and journal titles – not surprisingly there is some drift from the projections of their associated articles. But what we see in Figure 8.6 is that the general, relative proximities remain.

### 8.3.5    Abstracts as Proxies for Document Content

The numbers of nonunique words in the abstracts of the 5 articles are: 147, 153, 146, 232, and 212.

From Figure 8.7 it is seen that the abstracts also track very well the 5 selected articles. We find the reference sections to perform slightly better (cf. Figure 8.6 relative to Figure 8.7) in this tracking. On the other hand, as noted, the sizes of the abstracts are really quite small.

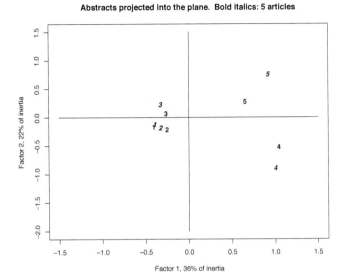

**FIGURE 8.7**
The Correspondence Analysis principal factor plane with the abstracts of all 5 articles projected onto the plane. For the analysis they are "passive" in that this projection takes place after the factors have been determined.

### 8.3.6 Conclusions on Study of Set of Articles

Consider the longer-term trajectory of the exploration and mapping of narrative from film-scripts.

The feasibility of using statistical learning methods in order to map a feature space characterization of film scripts (using two dozen characteristics) onto box office profitability was pursued by Eliashberg (Eliashberg et al. 2007 [110]). The importance of such machine learning of what constitutes a good quality and/or potentially profitable film script has been the basis for (successful) commercial initiatives, as described in (Gladwell 2006 [144]). The business side of the movie business is elaborated on in some depth in (Eliashberg et al. 2006 [109]). Our work complements the machine learning work of Eliashberg and his colleagues at the Wharton School, University of Pennsylvania, with data mining approaches.

The machine learning approach attempts to map filmscript inputs to target outputs. In practice, there is considerable manual intervention involved in data preprocessing, in order to find good characterization – features – for the film scripts. An important part of our data mining approach attempts to find faint patterns and structure in the filmscript data that can be exploited for a range of decision making, including such learning of mapping onto box-office success.

Having considered a feasible objective of filmscript-based narrative analysis, let us now consider the same objective applied to the research literature.

Just as for filmscripts, research funding agencies receive proposals and evaluate them according to various criteria. The evaluation usually is based on peer review, where the peer evaluators assess the fit of the proposal to the evaluation criteria. A number of phases of the evaluation process may be relevant, and also a number of phases in the decision-making process. The outcome is a decision to fund to a certain level or to decline.

Let us go further. The evaluation criteria are tied to output performance targets. Examples typically include research papers published, business initiatives, PhDs produced, and so on. From an input proposal with textual and numeric data the whole process is a mapping onto target outputs. Whether this process can be largely automated is an interesting issue to raise. We have seen how this issue has been raised and actively pursued in the filmscript context.

One can envisage similar application to discipline mapping and to individual-oriented performance mapping. National and international performance reviews are now *de rigueur* in research. The narrative that accompanies the evolution of a discipline, or of an individual's work, is one way to assess performance.

## 8.4   Conclusions

We have reported on how the structure of narrative, as expressed by a filmscript, can be analyzed. Then we turned attention to the analysis of the research literature. Questions that we are starting to address include the following. What is the narrative thread that connects one person's research outputs? What is the narrative thread in successive, over time, funding proposals? Targeted therefore is the researcher or team of researchers; funding agencies; and policy makers at the national level or internationally. See (Murtagh 2010 [237]) for first results in this direction.

# 9

## Measuring Classifier Performance: On the Incoherence of the Area under the ROC Curve and What to Do about It

David J. Hand

## CONTENTS

## 9.1   Introduction

Supervised classification problems appear in many guises: in medical diagnosis, in epidemiological screening, in creditworthiness classification, in speech recognition, in fault and fraud detection, in personnel classification, and in a host of other applications. Such problems have the same abstract structure: given a set of objects, each of which has a known class membership and for each of which a descriptive set of measurements is given, construct a rule that will allow one to assign a new object to a class solely on the basis of its descriptive measurement vector. Because such problems are so widespread, they have been investigated by several different (though overlapping) intellectual communities, including statistics, pattern recognition, machine learning, and data mining, and a large number of different techniques have been developed (see, for example, Hand 1997 [154], Hastie et al. 2001 [162], and Webb 2002 [314]). The existence of this large number of distinct approaches prompts the question of how to choose between them. That is, given a particular classification problem, which of the many possible tools should be adopted?

This question is answered by evaluating the tools, and choosing one that appears to perform well. To do this, of course, one needs to use a suitable criterion to assess each rule's performance. Unfortunately, since performance has many aspects, this leads to complications. At a high level, one might be interested in such things as how quickly a rule can be constructed, how quickly it can produce classifications, whether it can handle large datasets or very high numbers of descriptive characteristics, how well it copes with missing data, and so on. At a more immediate level, one will generally be interested in the extent to which the rule assigns new objects to the correct class: the *classification accuracy*. This is a general term, which encompasses a number of different measures, some of which we describe below.

Accuracy measures are also important for the design of classifiers. For example, building classifiers typically requires one to choose predictive variables and estimate parameters. One can think of these as exercises in classifier comparison. When deciding whether to include an additional variable $x$, one implicitly (sometimes explicitly) compares performance with and without this variable. When estimating parameters, one is implicitly comparing classifiers built using different values of the parameter. For all such exercises, some measure of performance is needed, so that a choice can be made between the different alternatives.

This chapter is concerned with such measures of classification accuracy classification. In particular, it focuses on one especially widely used measure – the *Area Under the ROC Curve* (AUC) – and variants of it such as the partial AUC (PAUC) and Gini coefficient, pointing out that all of these measures have a fundamental flaw. This is serious, because it means that mistaken conclusions can be drawn: that a poor classifier, diagnostic method, fault detector, etc., could be chosen in preference to a superior one if one of these measures is chosen. This problem is all the more serious because it has not been widely recognized.

Several books have now appeared that focus on the topic of evaluating classification methods. They include (Hand 1997 [154]) (which concentrates on classification accuracy and probability estimate measures), (Egan 1975 [105]), (Gönen 2007 [145]), and (Krzanowski & Hand 2009 [191]) (all three of which focus on ROC curves), and (Pepe 2000 [255]) and (Zhou et al. 2002 [321]) (both of which are devoted to evaluating classification accuracy measures within medicine).

For convenience, in this chapter we will assume that there are just two classes, which we will label as class 0 and class 1.

Section 9.2 sets the scene, Section 9.3 defines the *AUC*, Section 9.4 demonstrates its fundamental incoherence, Section 9.5 describes an alternative performance criterion which sidesteps the problem, and Section 9.6 discusses the wider context.

---

## 9.2    Background

A classification rule consists of a function mapping the descriptive vectors of objects to scores, $s$, along with a classification threshold, $t$, such that objects with scores $s \succ t$ are assigned to class 1, and otherwise to class 0. Unfortunately, classification rules are seldom perfect: the complications of real problems, and the limitations of real data, mean that classification rules typically assign some objects to the incorrect class. Different rules, with different mapping functions and different choices of $t$, misclassify different objects and different numbers of objects, and the aim of performance assessment is to measure the extent of misclassification.

To assess the performance of a rule, it is applied to a set of objects each of which have a known true class membership. For well-understood reasons, this set of objects should not be the same as the set used to construct the rule (that is, to estimate its parameters, choose the predictive variables, etc.), since performance assessment based on this set is likely to lead to an optimistic estimate of future performance. This problem can be tackled by using an independent "test set" of objects, or by using sophisticated resampling methods such as cross-validation, leave-one-out, or bootstrap methods – see, for example, (Hand 1997 [154]) for a discussion. In this chapter we assume that an appropriate such method has been applied, and we will not discuss it further.

Applying the rule to the set of objects used for evaluation yields a set of scores for the class 0 objects and a set of scores for the class 1 objects. We thus have empirical score distributions for each of the classes. In what follows, we will assume that the evaluation set is large enough that we can ignore sampling variability, and treat these two score distributions,

$f_0(s)$ and $f_1(s)$ as the true score distributions for the two classes. If one wished to construct confidence intervals, or make significance or hypothesis tests, then one would need to take account of the fact that the empirical distributions are in fact merely estimates of the true distributions $f_0(s)$ and $f_1(s)$.

When these distributions of scores are compared with threshold $t$, a proportion $1 - F_0(t)$ of the class 0 objects are misclassified, and a proportion $F_1(t)$ of the class 1 objects are misclassified, where $F_0(t)$ and $F_1(t)$ are the the cumulative distributions corresponding to $f_0(s)$ and $f_1(s)$. Each of these proportions is an indicator of the effectiveness of the classifier, but each, by itself, clearly fails to capture the full details of the performance. In particular, by increasing $t$, one can make the misclassified proportion of class 0 objects as small as one likes, though at the cost of increasing the proportion of class 1 objects that are misclassified. And vice versa. Some way of combining these two misclassification proportions is required in order to yield an overall measure of performance.

If one regards class 0 as the class of particular interest (which we shall call "cases," meaning, for example, sick people, fraudulent credit card transactions, faulty devices, etc.) then the proportion of class 0 correctly classified, $F_0(t)$, is often called the sensitivity, especially in medical and epidemiological circles. The corresponding proportion of class 1 correctly classified, $1 - F_1(t)$, is often called the specificity.

These are both proportions conditional on the known true classes. Similar proportions conditional on the predictions can be defined: the positive predictive value is the proportion of those objects predicted to be cases that really are cases. The negative predictive value is the proportion of those objects predicted to be noncases, which really are noncases.

If a proportion $\pi$ of the objects in the entire population belong to class 0 (and hence a proportion $1 - \pi$ to class 1) then we can define the misclassification rate, or error rate, $e$, as the overall proportion of objects that the classifier misclassifies. Note that we have

$$e(t) = \pi(1 - F_0(t)) + (1 - \pi)F_1(t)$$

The complement of this, $p = 1 - e$, is of course the proportion correctly classified. Error rate is the single most widely used performance measure in the pattern recognition and machine learning literature (Jamain 2004, Jamain & Hand 2008 [176, 177]).

This illustrates a fundamental point about two-class classification problems. There are two kinds of misclassification, and if we want to define a single numerical performance measure that can then be used to compare classifiers then we need to find some way to combine these two kinds. More generally, we can say that there are two degrees of freedom, since the proportions incorrect for the two classes are just one possible way of parameterizing things.

A characteristic of the error rate $e$ is that it regards the two kinds of misclassification as equally serious in its defining weighted sum. In fact this is very rarely appropriate, and it is far more usual that one kind is more serious than the other – though it may be very difficult to say just how much more serious it is. This observation drives much of the later development in this chapter.

If we relax the assumption of equal misclassification severity, and generalize things by letting misclassifications of class 0 objects have severity ("cost") $c_0$ and misclassifications of class 1 objects have severity $c_1$, then the overall loss due to misclassifications, when using a threshold $t$, is

$$Q(t; c_0, c_1) = \pi c_0 \left((1 - F_0(t)) + (1 - \pi)c_1 F_1(t)\right.$$

We have seen that error rate is the special case of $Q$, which sets $c_0 = c_1 = 1$. Another important special case is the Kolmogorov-Smirnov statistic between the two distributions

$f_0(s)$ and $f_1(s)$, which is obtained by setting $(c_0, c_1) \propto (\pi^{-1}, (1-\pi)^{-1})$. This is widely used in the credit scoring industry in the United States.

Another variant of $e$, typically written in terms of the proportion correct rather than incorrect, standardizes it by the proportion one would expect to get right by chance if the classifier had no predictive power. This is the kappa statistic. The expected proportion correct under the "no predictive power" assumption, $P$, is obtained simply by multiplying the marginals of a cross-classification of true class by predicted class, and adding the two cells corresponding to correct classifications. The kappa statistic is then

$$\kappa = (p - P)(1 - P)$$

Another way of combining the two degrees of freedom, which is popular in the information retrieval community, is the $F$ measure, defined as the harmonic mean of sensitivity and positive predictive value. As this perhaps indicates, there is an unlimited number of other ways that the two degrees of freedom implicit in the problem can be combined to yield a single performance measure. Each measure will have different properties, and will tap into a different aspect of classifier performance: there is no "right" measure. Ideally a measure should be chosen that best reflects those aspects of the problem that are of primary concern. This, however, is both difficult and time consuming, so researchers generally adopt some standard measure – hence, the widespread use of error rate, despite it usually being inappropriate.

There is another, rather different class of measures that should also be mentioned. For many classification methods the score is an actual estimate of the probability that an object belongs to class 0 (or class 1), and this is then compared to a threshold. For others, which just yield a score that (one hopes) is monotonically related to this probability, one can transform it into such a probability by a calibration exercise. For some problems it is appealing to know how accurate are the probabilities given by the classifier: that, for example, 80% of the objects, which a classifier estimates as having a 0.8 probability of belonging to class 0, really do belong to class 0 – or how far this value is from 80%. There are many examples of such probability scores, including the Brier score and log score (see, e.g., Hand 1994 [153]). However, as Friedman demonstrates in an elegant paper (Friedman 1997 [136]), "more accurate probability estimates do not necessarily lead to better classification performance and often can make it worse." He added that "these results suggest that in situations where the goal is accurate classification, focusing on improved probability estimation may be misguided."

Apart from the probability scores mentioned in the preceding paragraph, all of the criteria mentioned above hinge on a value for the threshold $t$ having been chosen. As $t$ is varied over the range of the scores, so different values of $F_0(t)$ and $F_1(t)$ will be generated, so that different values of the criteria will result. The value of $t$ is thus critical in evaluating performance. The implications of this are described in the next section.

## 9.3   The Area under the Curve

A popular way of representing the fact that by varying $t$ one generates different values of the pair $F_0(t)$ and $F_1(t)$ is by means of a receiver operating characteristic curve, or ROC curve. This is a plot of $F_0(t)$ on the vertical axis against $F_1(t)$ on the horizontal axis (though different axes are sometimes used). A given classifier then traces out a curve from point $(0, 0)$ to point $(1, 1)$ as $t$ increases. Such a curve thus shows the performance of the classifier, in terms of the two kinds of misclassification, for any choice of threshold $t$. It follows immediately from the definition, that ROC curves are monotonically nondecreasing from $(0,0)$ to $(1,1)$. Moreover, perfect separation between the two distributions, so that perfect classification into the two classes could be achieved by an appropriate choice of $t$, corresponds to a ROC curve that passes through point $(0,1)$. In general, for any given value of $F_1(t)$ on the horizontal axis, a larger value of $F_0(t)$ corresponds to a better classification rule. This means that superior rules have ROC curves that lie nearer to the top left point $(0,1)$ of the ROC square. This observation has led to a compact summary of classification rule performance: the area under the ROC curve, or *AUC*. The larger is the *AUC* the better is the rule.

The AUC also has other attractive interpretations. It is the average sensitivity, assuming that each value of specificity is equally likely. It is also the probability that a randomly chosen object from class 0 will have a score lower than that of a randomly chosen object from class 1. This is the test statistic used in the Mann-Whitney-Wilcoxon nonparametric test to compare two distributions. This interpretation, though widely cited, is not all that helpful for understanding classification problems since the circumstances (one randomly drawn member from each class) seldom arise. An alternative and more useful interpretation is that the AUC is a simple linear transformation of the probability of correctly classifying randomly chosen objects using a classifier that assigns objects to classes by comparing scores with a threshold, which is itself randomly drawn from the mixture distribution of the scores of the two classes (Hand 2010 [158]).

An outline of ROC curves, their properties, and applications, is given in (Krzanowski & Hand 2009 [191]).

## 9.4 Incoherence of the Area under the Curve

For convenience of exposition, in what follows, we will assume that $F_0(t)$ stochastically dominates $F_1(t)$, that $F_0(t)$ and $F_1(t)$ are everywhere differentiable, and that the ratio $f_0(t)/f_1(t)$ is monotonically decreasing. These assumptions do not impose any constraints on the theoretical development, but merely allow us to ignore anomalies. A full discussion of how to handle general situations when they do not apply is given in (Hand 2009 [156]).

We saw above that the cost weighted misclassification rate – the overall loss – when using a classification threshold $t$ was

$$Q(t; c_0, c_1) = \pi c_0 \left( (1 - F_0(t)) + (1 - \pi)c_1 F_1(t) \right) \tag{9.1}$$

This loss is a function of the specified cost pair $(c_0, c_1)$ and also of $t$.

For a given cost pair, it is clearly rational to choose $t$ so as to minimize the loss – any other choice would incur more misclassification loss than was necessary. Sometimes other, problem specific, issues may force choices that do not minimize the overall loss. For example, if the classification is an element of a screening program, in which those predicted to be cases are to be subjected to a closer and more expensive investigation, and where the overall resources for this more expensive investigation are limited, then one might be forced to choose $t$ so that only a small proportion are predicted to be cases. In this discussion, however, we assume that the choice will be made that minimizes the overall loss.

We denote this loss-minimizing choice of classification threshold by $T$. By virtue of the mathematical assumptions on $f_0(s)$ and $f_1(s)$, we can apply a little calculus to (9.1) to reveal that

$$\pi c_0 f_0(T) = (1 - \pi)c_1 f_1(T), \tag{9.2}$$

and hence an invertible relationship (recall our assumption that the ratio $f_0(t)/f_1(t)$ was monotonic) between the ratio $r = c_0/c_1$ and the classification threshold $T$:

$$r = \frac{c_0}{c_1} = \frac{(1-\pi)}{\pi}\frac{f_1(T)}{f_0(T)}. \tag{9.3}$$

This is all very well if one can decide what cost pair $(c_0, c_1)$ is appropriate. Often, however, this is difficult. For example, a typical situation arises when the classification rule is to be applied in the future, in circumstances that cannot be entirely predicted. A medical diagnostic system may be applied to different populations, in different clinics or even in different countries, where the consequences of the two different types of misclassification vary from population to population. A credit scoring classifier, predicting who is likely to default on a loan and who is not, may have a cost pair that varies according to the economic climate. A face recognition system for detecting terrorist suspects at airports may have misclassification costs that depend on the level of threat. And so on. And changing costs mean a changing optimal ratio $T$.

If the costs are difficult to determine, one might be able to contemplate distributions of likely future values for the ratio of the pair $(c_0, c_1)$, and hence of $T$ (related via (9.3)). In this case, a suitable measure of performance is given by integrating $Q(T(c_0, c_1); c_0, c_1)$ over the distribution of $(c_0, c_1)$ values, where we have explicitly noted the dependence of the optimal $T$ on $(c_0, c_1)$. If we let $g(c_0, c_1)$ be the joint distribution of the two misclassification costs, then this performance measure is

$$L = \int\int \{\pi c_0 (1 - F_0(T(c_0, c_1))) + (1-\pi)c_1 F_1(T(c_0, c_1))\} g(c_0, c_1)dc_0 dc_1. \tag{9.4}$$

In fact, however, we have seen from (9.3) that $T$ depends only on the ratio $r = c_0/c_1$, so we can rewrite (9.4) as

$$L = \int\int \{\pi r c_1 (1 - F_0(T(r))) + (1-\pi)c_1 F_1(T(r))\} h(r, c_1)dr dc_1, \tag{9.5}$$

where $h$ includes the Jacobian of the transformation from $(c_0, c_1)$ to $(r, c_1)$.

Equation (9.5) can be rewritten as

$$\begin{aligned}
L &= \int \{\pi r (1 - F_0(T(r))) + (1-\pi)F_1(T(r))\} \int c_1 h(r, c_1)dc_1 dr \\
&= \int \{\pi r(T)(1 - F_0(T)) + (1-\pi)F_1(T)\} w(T)dT \\
&= (1-\pi)\int \{f_1(T)(1 - F_0(T)) + f_0(T)F_1(T)\} \frac{w(T)}{f_0(T)}dT
\end{aligned} \tag{9.6}$$

where $w(T)$ includes $\int c_1 h(r, c_1)dc_1$ and the Jacobian of the transformation from $r$ to $T$, and where the last line has used (9.3).

Suppose, now, that we choose function $w(T)$ such that $w(T) = f_0(T)$. Then (9.6) becomes

$$L = (1-\pi)2\left\{1 - \int F_0(T)f_1(T)dt\right\}$$

Now, $\int F_0(T)f_1(T)dt$ is the area under the ROC curve, the AUC, so that the loss in (9.6) can be expressed as

$$L = 2(1-\pi)\{1 - AUC\}$$

and

$$AUC = \left( 1 - \frac{L}{2(1 - \pi)} \right).$$

What this shows is that the AUC is linearly related to the expected misclassification loss if the classification threshold $T$ is randomly chosen according to the distribution $f_0(T)$.

Now, equation (9.3) shows that the relationship between $T$ and $r$ involves the two distribution functions $f_0$ and $f_1$. This means that if we choose $T$ from a given distribution $w(T)$, then the distribution of $r$ will depend on the functions $f_0$ and $f_1$ – the function of $r$, which is used in the definition of the loss $L$ in (9.5), will depend on $f_0$ and $f_1$. Since these functions will vary from classifier to classifier, so will the function of $r$. In particular, if we choose $w(T) = f_0(T)$, we find that the AUC is equivalent to the overall loss, but a loss that uses different distributions over the relative cost severity for different classifiers. This seems untenable: the choice of a likely distribution for the misclassification cost ratio cannot depend on the classifier being used, but has to be based on considerations distinct from these classifiers. Letting the distribution of $r$ depend on the empirical distributions is equivalent to saying that misclassifying someone with H1N1 virus as healthy (so that they are not treated) is ten times as serious as the converse if (say) logistic regression is used, but is a hundred times as serious if a tree classifier is used. But the relative severity of misclassifications cannot depend on the tools one happened to choose to make the classifications.

## 9.5 What to Do about It

The problem with the AUC described in the previous section arises because the misclassification loss is averaged over a distribution of the misclassification cost ratio that depends on the score distributions for the two classes. To overcome this, we need to average over a distribution that is independent of these score distributions. Of course, such a distribution could be chosen in an unlimited number of ways. Different ways correspond to different aspects of the classifier's performance. There is no way that this intrinsic arbitrariness in the performance metric can or should be eliminated – different researchers may be interested in different aspects of performance. However, if we do choose a specific distribution, and one which is independent of the score distributions, then the comparisons between classifiers will be based on the same metric. The key point is that, while it is perfectly reasonable, even desirable, that different researchers may choose different metrics, it is unreasonable that the same researcher should choose different metrics for different classifiers if their aim is to make a statement about comparative performance.

With these points in mind, there are two classes of approach to choosing distribution for $r$, and we recommend they are used in parallel.

The first class is based on the fact that often, while one might find it difficult to come up with a global distribution for the different values of the cost ratio, one is likely to know something about it. For example, in the H1N1 virus example above, one might feel that misclassifying a sick person as healthy (so that they are not treated) is more serious than the reverse (when a healthy person is unnecessarily treated). This would mean (taking class 1 as healthy) that $c_1 \prec c_0$, so that $r$ was more likely to take values greater than 1. This can be reflected in the choice of distribution for $r$.

By definition, the cost ratio $r$ treats the two kinds of misclassifications asymmetrically. For convenience, (Hand 2009 [156]) transforms $r$ to $c = (1 + r^{-1})$. $c$ lies between 0 and 1, and takes value $1/2$ when $c_0 = c_1$. (Hand 2009 [156]) then suggests using a beta distribution for $c$:

$$\nu(c) = beta(c; \alpha, \beta) = c^{\alpha-1}(1-c)^{\beta-1}/B(1; \alpha, \beta).$$

where $B$ is the beta function. If the parameters $\alpha$ and $\beta$ are both greater than 1, such a distribution takes its mode at $(\alpha-1)(\alpha+\beta-2)$ and this can be chosen to match the most likely value for $c$ (and hence for $r$). Since different researchers may wish to place different degrees of probability on different values of $r$, they would quite properly choose different beta distributions, and they would again quite properly obtain different performance measurements for the same classifier.

The second approach is to adopt a universal standard for the cost ratio distribution so that all researchers are using the same performance criterion. (Hand 2009 [156]) suggests that the $beta(c; 2, 2)$ distribution is used. This is symmetric about $c = 1/2$, and decays to 0 at $c = 0$ and $c = 1$.

There is an analogy here to the use of error rate. This is a universal standard, based on the choice $(c_0, c_1) = (1, 1)$, so all researchers using error rate are using the same performance criterion. However, such a choice may not properly reflect a particular researcher's cost values. In that case he or she can adopt whatever values are felt appropriate, to yield a performance criterion specific to them.

In summary, if something is known, or can be asserted, about the likely values of $c$ (or $r$) then an appropriate beta distribution should be used. Each researcher can then choose between classification rules on the basis of what they believe are the likely relative severities of the misclassification costs. However, to facilitate communication, and to make the performance clear to other researchers, we recommend that the standard $beta(c; 2, 2)$ approach is also always reported.

The AUC takes values between 0 and 1, with large values indicating good performance. For interpretative convenience we propose applying a simple transformation to the average loss, dividing by its maximum value and subtracting from 1, so that it, too, lies between 0 and 1, with large values meaning good performance. This yields the $H(\alpha, \beta)$ performance measure. An $R$ routine for evaluating $H(\alpha, \beta)$ is available on http://www.ma.ic.ac.uk/~djhand.

## 9.6    Discussion

The AUC is a very widely used measure of classification rule performance. It also appears in other guises and variations. For example, a chance-standardized version, the so-called Gini coefficient, subtracts as a baseline the AUC expected from a ("chance") classifier, which has no predictive power (and therefore that has $AUC = 1/2$) and divides by the difference between the AUC of the best conceivable classifier ($AUC = 1$) and this baseline AUC:

$$Gini = (AUC - 1/2) / (1 - 1/2)) = 2 \cdot AUC - 1.$$

Because the Gini coefficient is a simple linear transformation of the AUC it is subject to exactly the same criticisms.

If one thinks of the AUC, $\int F_0(s)f_1(s)ds = \int F_0()dF_1(s)$, as the average value of $F_0(s)$ integrated over a uniform distribution over $F_1(s)$ then one can immediately see that in some cases only certain ranges of $F_1(s)$ are likely to be relevant. The Partial AUC (PAUC) addresses this by restricting the range of the integral. But this does not remove the fundamental fact that such an integral, by virtue of the relationship between threshold and cost ratio, implies that different classifiers are implicitly using different distributions over the relative misclassification severities: the PAUC is subject to the same criticisms as the AUC.

The particular merit of the AUC is that it compares classifiers without requiring a threshold value to be chosen. It does this by implicitly integrating over a distribution of possible threshold values. That in itself is not problematic, but if one believes that the threshold is most appropriately chosen to minimize the overall loss arising from misclassifications then the particular distribution implicit in the definition of the AUC is equivalent to supposing that different classifiers use different distributions over the possible values of relative misclassification costs. This is surely incorrect: the costs one incurs from making an incorrect classification depend only on the classification, not by what means it was reached. (We have, of course, ignored issues such as the computational costs in applying the classifiers).

The intrinsic arbitrariness of the distribution of relative misclassification costs cannot be removed. Different such distributions reflect different aspects of performance. In particular, different researchers, or even the same researcher at different times, might be interested in different aspects of performance, and so might wish to use different relative cost distributions. However, for a single researcher to use different distributions for different classifiers on exactly the same problem is wrong. It means one is evaluating classifiers using a different evaluation criterion for each classifier. This is exactly what the AUC does. It is overcome by fixing the relative misclassification cost distribution, as, for example, in the H-measure.

In some contexts, considerations other than overall misclassification loss might lead one to choose a classification threshold that does not minimize this loss. For example, cost considerations might lead one to fix the proportion of the overall population that one can accept as "predicted cases," or the proportion of either one of the classes that one can accept as "predicted cases." If one was uneasy about specifying a particular such proportion to be used in the future, one would adopt a distribution of likely proportions. This distribution would then correspond to a distribution over the threshold $t$. Unfortunately, the fundamental complementarity between the cost ratio and the proportion below the threshold, implicit in (9.3), means that one cannot choose both a distribution over the proportion accepted and the relative cost distribution. One must choose one or the other. Choosing the proportion accepted means one is choosing an operating characteristic of the classifier (ROC curves are not called receiver operating characteristic curves by accident). Choosing a distribution over the relative costs is saying what you believe applies to the problem. The latter seems to dominate the former.

Further discussion of these issues are given in (Hand 2009, 2010 [156], [157]), and further illustrations are given in (Hand 2008 [155]) (breast cancer diagnosis through proteomic analysis) and (Hand & Zhou 2010 [159]) (classifying customers who are in default with credit repayments).

# 10

# A Clustering Approach to Monitor System Working: An Application to Electric Power Production

**Alzennyr Da Silva, Yves Lechevallier, and Redouane Seraoui**

## CONTENTS

## 10.1   Introduction

Clustering is one of the most popular tasks in knowledge discovery and is applied in various domains (e.g., data mining, pattern recognition, computer vision, etc.). Clustering methods seek to organize a set of items into clusters such that items within a given cluster have a high degree of similarity, whereas items belonging to different clusters have a high degree of dissimilarity. Partitioning clustering methods ([114, 148, 175]) aim to obtain a single partition of the input data into a fixed number of clusters. Such methods often look for a partition that optimizes (usually locally) an adequacy criterion function. Clustering techniques have been widely studied across several disciplines, but only a few of the techniques developed scale to support clustering of very large time-changing data streams (Madjid & Norwati 2010 [220]) (Mahdiraji 2009 [221]). The major challenge in clustering of evolving data is to handle cluster evolution: new clusters may appear, old ones may disappear, merge or split over time.

In this chapter, we propose a clustering approach based on nonoverlapping windows to monitor system working and detect changes on evolving data. In order to measure the level of changes, our approach uses two external evaluation indices based on the clustering extension. In order to deal with the cluster follow-up problem, the proposed approach uses the clustering extension, i.e., a rough membership function, to match clusters in subsequent time periods.

This chapter is organized as follows. Section 10.2 presents related work. Section 10.3 presents our clustering approach to monitor evolving data streams. Section 10.4 describes the experiments carried out on real data from electric power production. The conclusion and future work are presented in Section 10.5.

## 10.2   Related Work

Over the past few years, a number of clustering algorithms for data streams have been put forward (Madjid & Norwati 2010 [220]). Five different algorithms for clustering data streams are surveyed in (Mahdiraji 2009 [221]).

One of the earliest and best known clustering algorithms for data streams is BIRCH (Zhang et al. [320]). BIRCH is a heuristic that computes a preclustering of the data into so-called clustering feature vectors (micro clustering) and then clusters this preclustering using an agglomerative bottom-up technique (macro clustering). BIRCH does not perform well because it uses the notion of radius or diameter to control the boundary of a cluster. STREAM (O'Callaghan et al. 2001 [248]) is the next main method that has been designed specially for data stream clustering. This method processes data streams in batches of points represented by weighted centroids that are recursively clustered until $k$ clusters are formed. The main limitation of BIRCH and STREAM is that old data points are equally important as new data points.

CluStream (Aggarwal et al. 2003 [2]) is based on BIRCH. It uses the clustering feature vectors to represent clusters and expands on this concept by adding temporal features. This method takes snapshots at different timestamps, favoring the most recent data. A snapshot corresponds to $q$ (where $q$ depends on the memory available) micro-clusters stored at particular moments in the stream. One weakness of the algorithm is the high number of parameters that depend on the nature of the stream and the arrival speed of elements.

StreamSamp (Csernel et al. 2006 [78]) is based on CluStream and combines a memory-based reduction method and a tilted window-based reduction module. In StreamSamp, summaries with a constant size are maintained. The summaries cover time periods of varying sizes: shorter for the present and longer for the distant past. StreamSamp has the advantage of being fast on building the data stream summary. However, its accuracy degrades over time because old elements increase in weight for a given sample size.

TECNO-STREAMS (Nasraoui et al. 2003 [240]) is an immune system inspired single pass method to cluster noisy data streams. The system continuously learns and adapts to new incoming patterns. In (Nasraoui et al. 2008 [239]), the authors extended their work to the categorical domain. They present a case study on tracking evolving topic trends in textual stream data including an external data describing an ontology of the web content.

MONIC (Spiliopoulou et al. 2006 [286]) is a framework for modeling and tracking cluster transitions over time. In this framework, an offline clustering is applied periodically on an accumulating dataset and a weighting based scheme is used for processing past data. As the framework operates over an accumulating dataset and no strategy for summarizing data is applied, the performance of the system can decay. Moreover, if the accumulating dataset assumes huge dimensions it may not fit the available physical memory.

## 10.3 Clustering Approach for Monitoring System Working

The clustering process is started by splitting the input data stream into batches of equal size. A data batch is also referenced as a *window*.

Individuals within a batch are chronologically sorted according to their arrival date. Let $H^{(1)}$ be the batch containing the first incoming data. Let's also consider $H^{(2)}$ the next batch containing newer data over time. $H^{(1)}$ and $H^{(2)}$ are thus sequential batches with no overlapping area.

The next step consists of applying a clustering algorithm on data of each batch. Let $R^{(t-1)} = \{r_1^{(t-1)}, \ldots, r_c^{(t-1)}, \ldots, r_k^{(t-1)}\}$ be the set of representatives (prototypes) of each cluster obtained by the clustering on batch $H^{(t-1)}$. The next step consists of assigning data in batch $H^{(t)}$ to the prototypes of $R^{(t-1)}$, that defines a first partition $P_1^{(t)}$ containing $k^{(t-1)}$ clusters. After this, we apply the same clustering algorithm on data in $H^{(t)}$ defining a second partition $P_2^{(t)}$ with $k^{(t)}$ clusters represented by the set of prototypes $R^{(t)} = \{r_1^{(t)}, \ldots, r_c^{(t)}, \ldots, r_k^{(t)}\}$. This process is repeated over all two-by-two batches. A schema of the approach described is illustrated in Figure 10.1. It is important to note that the proposed approach is totally independent of the clustering algorithm used. The only requirement is that the algorithm should furnish a prototype for each cluster.

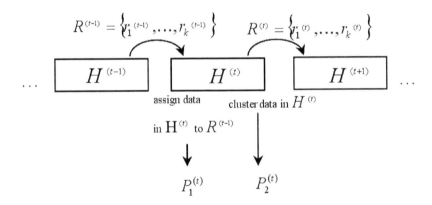

**FIGURE 10.1**
Schema of the clustering approach to monitor system working.

The main idea behind our change detection strategy is explained as follows. For a same data batch $H^{(t)}$, the partition $P_1^{(t)}$ is influenced by the clustering structure generated from data in the previous batch $H^{(t-1)}$. This partition will thus reflect the segmentation of the data in $H^{(t)}$ according to older patterns from $H^{(t-1)}$. On the other hand, the partition $P_2^{(t)}$ receives no influence from the past and only takes into account individuals in batch $H^{(t)}$. This partition will thus reflect the organization of current data. The occurrence of changes between two subsequent data batches can be detected by comparing partitions $P_1^{(t)}$ and $P_2^{(t)}$. In other words, if a change takes place, it will be caught by the degree of disagreement between two partitions from a same data batch.

### 10.3.1 Partition-Comparison Criteria

The entry $n_{ij}$ in each cell of Table 10.1 represents the number of individuals that are in both cluster $i$ of $P_1^{(t)}$ and cluster $j$ of $P_2^{(t)}$. The term $n_{i.}$ is the number of individuals in cluster $i$ of $P_1^{(t)}$ and $n_{.j}$ is the number of individuals in cluster $j$ of $P_1^{(t)}$.

**TABLE 10.1**

Confusion matrix between partitions $P_1^{(t)}$ and $P_2^{(t)}$.

| clusters of $P_1^{(t)}$ | clusters of $P_2^{(t)}$ | | | | | |
|---|---|---|---|---|---|---|
| | 1 | $\cdots$ | $j$ | $\cdots$ | $k^{(t)}$ | |
| 1 | $n_{11}$ | $\cdots$ | $n_{1j}$ | $\cdots$ | $n_{1k}$ | $n_{1.}$ |
| $\vdots$ | | | | | | |
| $i$ | $n_{i1}$ | $\cdots$ | $n_{ij}$ | $\cdots$ | $n_{ik}$ | $n_{i.}$ |
| $\vdots$ | | | | | | |
| $k^{(t-1)}$ | $n_{m1}$ | $\cdots$ | $n_{mj}$ | $\cdots$ | $n_{mk}$ | $n_{m.}$ |
| | $n_{.1}$ | | $n_{.j}$ | | $n_{.k}$ | $n_{..} = n$ |

The F-measure (van Rijsbergen 1979 [297]) is the harmonic mean of *Recall* and *Precision* (cf. equation 10.1). This measure seeks the best match of a cluster in partition $P_1^{(t)}$ for a cluster in partition $P_2^{(t)}$. The higher the values of *Recall* and *Precision*, the more similar are the clusters compared.

$$F = \sum_{i=1}^{k^{(t-1)}} \frac{n_{i.}}{n} max_{j=1,\ldots,k^{(t)}} \frac{2 Recall(i,j) Precision(i,j)}{Recall(i,j) + Precision(i,j)} \qquad (10.1)$$

where:

$$Recall(i,j) = \frac{n_{ij}}{n_{i.}} \text{ and } Precision(i,j) = \frac{n_{ij}}{n_{.j}}$$

The Corrected Rand index (Hubert & Arabie 1985 [171]) (cf. equation 10.2) – CR, for short – has its values corrected for chance agreement. The CR index applies a global analysis of the two partitions compared.

$$CR = \frac{\sum_{i=1}^{k^{(t-1)}} \sum_{j=1}^{k^{(t)}} \binom{n_{ij}}{2} - \binom{n}{2}^{-1} \sum_{i=1}^{k^{(t-1)}} \binom{n_{i.}}{2} \sum_{j=1}^{k^{(t)}} \binom{n_{.j}}{2}}{\frac{1}{2}[\sum_{i=1}^{k^{(t-1)}} \binom{n_{i.}}{2} + \sum_{j=1}^{k^{(t)}} \binom{n_{.j}}{2}] - \binom{n}{2}^{-1} \sum_{i=1}^{k^{(t-1)}} \binom{n_{i.}}{2} \sum_{j=1}^{k^{(t)}} \binom{n_{.j}}{2}} \qquad (10.2)$$

The F-measure takes a value in the range $[0, 1]$ and the CR index takes a value in the range $[-1, +1]$. For both indices, value 1 indicates a perfect agreement between the partitions compared, whereas values near 0 correspond to cluster agreements found by chance.

### 10.3.2 Cluster Follow-Up

Our approach incrementally matches clusters in subsequent data batches and characterizes their evolution over time. Cluster matches are verified by means of *Recall* (cf. equation 10.1).

A cluster survives if it conserves a minimum of its original individuals (*survivalThreshold*) from a data batch to the next one over time. A cluster splits if it conserves a minimum of its original individuals (*splitThreshold*) in the new clusters it originates in the next batch.

**Definition 4** *A cluster $i$ of $P_1^{(t)}$ survives if there is a cluster $j$ in $P_2^{(t)}$ such that $Recall(i, j) >= survivalThreshold$.*

**Definition 5** *The cluster $j$ of $P_2^{(t)}$ that matches the cluster $i$ of $P_1^{(t)}$ is the survivor cluster that conserves the maximum number of individuals of $i$, i.e., $\arg \max\limits_{j \in (1, \cdots, K^{(t)})} Recall(i, j)$.*

**Definition 6** *A cluster $i$ of $P_1^{(t)}$ splits if there are two or more clusters $j$ of $P_2^{(t)}$ such that $Recall(i, j) >= splitThreshold$.*

**Definition 7** *A cluster $i$ of $P_1^{(t)}$ disappears if it neither splits nor has a match in $P_2^{(t)}$.*

**Definition 8** *A cluster $i$ fuses with one or more cluster of $P_1^{(t)}$ when it matches a cluster $j$ of $P_2^{(t)}$ which is also a match for at least another cluster $i'$ of $P_1^{(t)}$, with $i' \neq i$.*

**Definition 9** *A cluster $j$ of $P_2^{(t)}$ is a new cluster if it is neither a match nor a son of a cluster of $P_1^{(t)}$.*

The clustering algorithm applied in our approach to automatically discover the number of clusters is described as follows. First, a self-organizing map (SOM) (Kohonen 1995 [189]) initialized by a principal component analysis (PCA) (Elemento 1999 [108]) is used to cluster data in each batch. The competitive layer of the SOM contains 200 neurons organized in a rectangular topology. Second, a hierarchical agglomerative clustering (HAC) using the Ward criterion is applied on the final prototypes obtained by the SOM. The resulting dendrogram is cut according to the elbow criterion (Aldenderfer & Blashfield 1984 [5]).

### 10.3.3  Cluster Description

In order to determine the most distinguishing variables of a cluster, we apply the cluster ratio, CR, measure (Celeux et al. 1989 [66]) defined according to equation 10.3.

$$CR(j, C) = \frac{B_c^j}{B_c} \text{ where } B_c^j = \mu_c d^2(r_c^j, r^j) \text{ and } B_c = \mu_c d^2(r_c, r) \tag{10.3}$$

The CR is based on the between-cluster inertia ($B$) of a partition with $k$ clusters. This measure computes the relative contribution of variable $j$ and cluster $C$ to the global between-cluster inertia. In equation 10.3, $B_c$ represents the contribution of cluster $C$ to the global between-cluster inertia and $B_c^j$ is the contribution of variable $j$ to $B_c$, $\mu_c$ is the weight of cluster $C$, $r_c$ is the representative (prototype) of cluster $C$, $r$ is the center of gravity of the data and $d$ is a dissimilarity or distance. Variables with greater values of CR are the ones that better describe the cluster.

## 10.4  Experiments

We analyzed data streams describing the working states of a nuclear power plant composed of 4 reactor coolant pumps over 5 years (from 2001 to 2005). Each pump is individually

**Leak rate of pump 3**

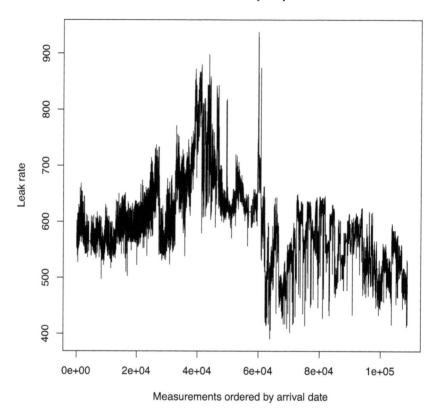

**FIGURE 10.2**
Leak rate of pump 3.

monitored by 10 sensors capturing different measurements (temperature, pressure, etc.). The global system (the reactor coolant circuit) is monitored by 7 sensors describing overall operation of the 4 pumps within the primary circuit. All sensors capture measurements at the same timestamp each 15 minutes. The data streams are then described by 10 local variables ($P_{ij}$ with $i \in \{1, \ldots, 4\}$ and $j \in \{1, \ldots, 10\}$ represents variable $j$ of pump $i$) describing each Reactor Coolant Pump and 7 global variables (referenced as $G_j$ with $j \in \{1, \ldots, 5\}$, $D$ and $T$) describing the Reactor Coolant Circuit.

We concentrated our experiments on data stream from pump 3 (chosen by the domain expert according to the leak rate parameter, cf. Figure 10.2) and applied a temporal partition of the data in time window with length equal to one month, $survivalThreshold = 70\%$ and $splitThreshold = 30\%$.

The values obtained by the two partition-comparison criteria (F-measure and CR index) are presented in the panels of Figure 10.3. The most instable periods are the ones corresponding to the lower values of the F-measure and CR index. Notice that changes are detected by the CR index in greater widths than the F-measure. The CR index provides a global measurement based on the whole set of clusters in the two partitions compared, it is thus more sensitive to changes.

The changes undergone by clusters throughout the analyzed months are summarized by

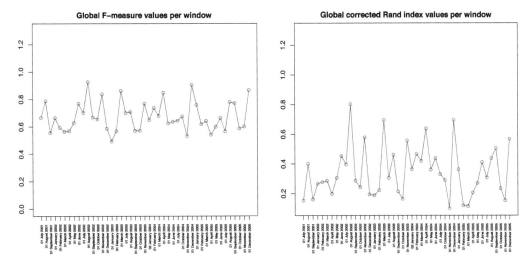

**FIGURE 10.3**
Values obtained by the F-measure (on the left) and by the CR index (on the right).

the graphics in Figure 10.4. The total number of clusters discovered by our approach varies between 3 and 8 (cf. panel *Total clusters* of Figure 10.4).

Let's consider consecutive months presenting great changes, for example *January 2003* compared to its previous month *December 2002* (cf. Figure 10.3: *f-measure=0.49* and *CR=0.18*). Indeed, both months present a total of 5 clusters (cf. panel *Total clusters* of Figure 10.4) and one would expect no changes to take place. However, looking in more detail at the transition occurring between these two months, our clustering approach reveals that clusters 1, 2, and 3 of *December 2002* have no correspondent cluster in *January 2003*, and they are thus considered as disappeared. Clusters 1, 2, and 4 of *January 2003* are new clusters. Finally, cluster 4 and 5 of *December 2002* match clusters 3 and 5 of *January 2003*, respectively. Details concerning the most significant variables and their prototype values within these months are given in Table 10.2.

Let's now consider an example of consecutive months presenting few changes, for example

**TABLE 10.2**
The most significant variables of clusters of December 2002 and January 2003.

| Month | Cluster | Most significant variables (and their values) |
|---|---|---|
| December 2002 | 1 | $P_{32}(62.3), T(13.8)$ |
| | 2 | $G_5(155.2), T(15.9)$ |
| | 3 | $P_{31}(37.4), P_{32}(64.9), T(16.9)$ |
| | 4 | $G_2(315.6), G_1(291.6)$ |
| | 5 | $D(4898.9)$ |
| January 2003 | 1 | $P_{31}(35.9), T(9.3), P_{32}(60.3)$ |
| | 2 | $G_4(178.8), P_{31}(37.5)$ |
| | 3 | $G_2(322.3), P_{32}(64), T(16.7)$ |
| | 4 | $G_4(175.15), G_3(30.3)$ |
| | 5 | $D(4594.8)$ |

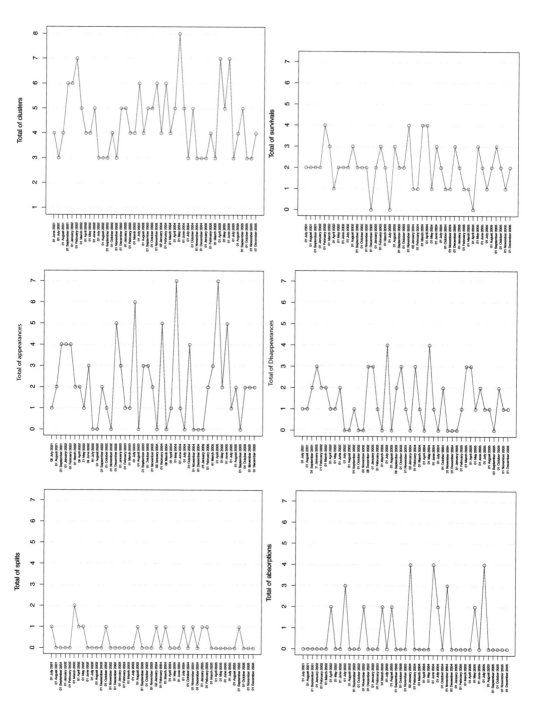

**FIGURE 10.4**
Summary of changes undergone by clusters over subsequent temporal data windows.

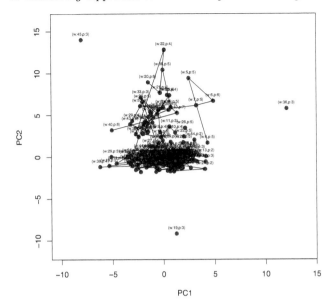

**FIGURE 10.5**
Projection of all cluster prototypes on the first factorial plane.

*August 2002* compared to its previous month *July 2002* (cf. Figure 10.3: *f-measure=0.92* and *CR=0.80*). Both months present exactly 3 clusters (cf. graphic *Total clusters* of Figure 10.4). Each cluster of *July 2002* matches a cluster in *August 2002*. No cluster disappearance, appearance, split, or fusions were verified. This means that the system working behaviors during these two consecutive months were very similar.

Figure 10.5 shows the projection of all cluster prototypes on the first factorial plane. In this figure, each circle represents a cluster prototype and arrows link a cluster prototype to its correspondent cluster prototype in the subsequent month. Circles with no input arrow represent new clusters and circles with no output arrow represent disappeared clusters. The legend $(w : a, p : b)$ represents cluster prototype $b$ in window $a$. Notice that most prototypes are concentrated around point $(0, 0)$. Changes of working states are represented by prototypes outside this central region. Special attention should be paid to the three marginal cluster prototypes $(w : 19, p : 3)$, $(w : 36, p : 3)$ and $(w : 45, p : 3)$. Each of these prototypes represents a particular cluster in *March 2003*, *February 2005*, and *November 2005* with a total of 5, 12, and 12 individuals, respectively. These are very thin clusters that would hardly be detected by a global analysis of the data. The most significant variable of cluster 3 in *March 2003* is $G_5$ and its value is equal to 157.7, whereas other cluster prototypes in this same month have value 155 for this variable. The most significant variables of cluster 3 in *February 2005* are $G_3$ and $P_{31}$ with respective values 32 and 40.7, whereas other cluster prototypes in the same month have values 37 and 44 for these variables. The most significant variables of cluster 3 in *November 2005* are $G_1$ and $G_2$ with respective values 294 and 306 on prototype, whereas other cluster prototypes in this same month have value 290 and 326 for these variables. In this context, Figure 10.6 illustrates individual memberships of cluster 3 in *November 2005*. In this figure, we can clearly notice 2 stable working states represented by clusters 1 and 2, whereas cluster 3 may represent a malfunctioning state.

In our approach, it is also possible to follow-up clusters over time. Let's consider cluster 5 from *January 2002*. According to the results given by our clustering

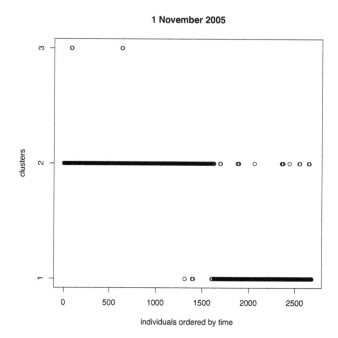

**FIGURE 10.6**
Memberships of individuals from November 2005.

approach, the cluster traceability and most significant variables (and their respective value on cluster prototype) can be described as follows. Cluster 5 (described by variables $G_3(42), G_1(293.7), G_2(308.8)$, and $P_{31}(48.3)$) in *January 2002* corresponds to cluster 6 (described by variables $G_3(41.7), P_{31}(48.3), G_1(292.7)$, and $G_2(312.4)$) in *February 2002* that corresponds to cluster 5 (described by variables $G_3(41.9), P_{31}(48.5), G_1(293.2)$, and $G_2(310.4)$) in *March 2002* that corresponds to cluster 2 (described by variables $G_2(310), G_1(293.2)$, and $G_3(40.5)$) in *April 2002* that corresponds to cluster 4 (described by variables $G_2(310.4), G_1(293.1)$, and $G_3(40.1)$) in *May 2002*, which disappeared in *June 2002*. In the e-commerce domain, for example, this functionality would be very useful to follow-up customer purchase behavior over time.

## 10.5   Conclusion

Condition monitoring is essential to the timely detection of abnormal operating conditions in critical systems. This issue is mostly addressed by developing a (typically empirical) model of the system behavior in normal conditions and then comparing the actual behavior of the system with the one predicted by the model. A deviation between the measured and predicted values of system signals can reveal a component malfunctioning. In model-based strategies, the construction of an accurate model representing the system normal operation is a nontrivial and important issue.

This chapter presents a clustering approach to monitor system working and detect changes over time. The main novelty in this approach concerns the change detection strategy that is based on the clustering extension, i.e., individual memberships reflecting the current system state. The changes are measured by the disagreement level between two partitions obtained from a same temporal data window. The proposed approach is positioned in a higher abstraction level of the clustering process and is totally independent of the clustering algorithm applied. It tracks cluster changes (splits, scissions, appearance, and disappearance) in subsequent windows and provides a snapshot of data evolution over time. Within this framework, nonincremental methods can be used in an incremental process which can be stopped and restarted at any time.

The proposed approach is generic and can be applied in different domains (e-commerce, web usage mining, industrial process monitoring, etc.) such as the one described in the experiments where a cluster represents a system working state composed of different sensor measurements.

## Acknowledgments

This work was supported by the ANR MIDAS project under grant number ANR-07-MDCO-008, CAPES-Brazil and INRIA-France.

# 11

## Introduction to Molecular Phylogeny

Mahendra Mariadassou and Avner Bar-Hen

## CONTENTS

## 11.1 The Context of Molecular Phylogeny

### 11.1.1 The Basis

The pioneering work of Charles Darwin (Darwin 1859 [82]), on which modern evolutionary biology has been built, has radically changed our understanding of evolution. Darwin introduced in his book *On the Origin of Species* a theory of evolution, whereby species evolve over generations through the selection process and natural diversity of living is obtained through the gradual accumulation of differences in subpopulations of species.

The evolution can be regarded as a branching process in which subpopulations of a species are transformed by accumulating differences before a species is separated from this parent species to form a new species or face extinction. The evolutionary tree illustrates the concept of evolution and the formation of new species from species already existing. Kinship ties that bind a group of species are commonly represented in the form of evolutionary trees, called "phylogenetic trees" or "phylogenies." An example of an evolutionary tree for primates is given in Figure 11.1.

All methods of reconstructing phylogenetic trees are based on the same intuitive idea: given that evolution occurs by the accumulation of differences, two species that diverged recently are "more" similar than two species whose divergence is older. The similarity between species was measured by morphological-type criteria (like the shape of bones, the number of legs, or the number of teeth) until the 1950s. The discovery of the structure of DNA by Watson and Crick in 1953 (Watson & Crick 1953 [312]) and the sequencing capabilities and analysis of macrobiological molecules that quickly followed have greatly changed the situation. Instead of establishing kinship from morphological criteria (the number is generally low and some are strongly subject to the discretion of the experimenter), we can now rely on phylogenetic molecular data, i.e., on DNA sequences (DNA) or protein. This revolution has three major advantages. First, evolution acts in a much finer way at the molecular level than do morphological characters: some mutations of the DNA sequence

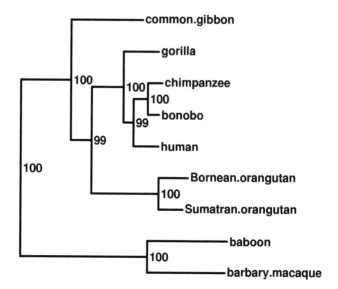

**FIGURE 11.1**
Evolutionary tree for primates with bootstrap probabilities for the nodes.

are invisible at the morphological level. Second, molecular sequences are less subject to the subjectivity of the experimenter than the morphological criteria. Finally, the sequences of molecular datasets provide far more important datasets than morphological criteria: instead of comparing species on some morphologic criteria (few tens), we compare sequences of many thousands of base pairs, or even several million for species whose entire genome is known.

Reconstructing the evolutionary history of species is obviously a goal in itself for evolutionary biologists. The most iconic project is the "Tree of Life" (Tree of Life Project, www.tolweb.com), which seeks to reconstruct the phylogenetic tree of all living species. Phylogenetic trees are also of major interest in other areas of biology. They are invaluable in comparative genomics, where they allow for example to predict the function of an unknown gene from the function of similar genes in closely related species (Eisen, and Eisen & Wu 1998, 2002 [106, 107]) or to predict, from the phylogenetic tree, whether two proteins interact or not (Pazos & Valencia 2001 [253]). But the scope of phylogeny cannot be reduced to molecular biology: phylogenies also appear naturally in conservation biology, particularly in studies measuring biodiversity (Bordewich et al. 2008 [32]).

## 11.2    Methods for Reconstructing Phylogenetic Trees

All applications described in Section 11.1.1 are based on well-reconstructed phylogenetic trees. But reconstructing such trees is a difficult task: we need to reconstruct the path followed by evolution from fingerprints it leaves on genomes, knowing that these fingerprints can be tenuous and fade anyway over time. Since Darwin, systematists rebuild evolutionary trees with an amazing precision.

There are essentially five major families of methods to reconstruct a phylogeny: methods of parsimony (Edwards & Cavalli-Sforza 1963 [103]), the least-squares methods (Cavalli-

Sforza & Edwards 1967 [65]), maximum likelihood methods ([104, 118]), distance methods (Fitch & Margoliash 1967 [129]), and Bayesian methods ([224, 206, 268]). The major contribution of the work of Cavalli-Sforza and Edwards, both disciples of Fisher, is probably early identification of the reconstruction of phylogenetic trees as a matter of statistical inference.

All methods mentioned above can be decomposed into three parts:

1. An *optimality criterion*, which measures the adequacy of data to a given phylogenetic tree (e.g., parsimony, likelihood, sums of squares, etc.);

2. A *search strategy* to identify the optimal tree (e.g., exhaustive search, gradient descent, etc.)

3. Assumptions on *mechanism of evolution data.*

There is no method superior to all others, each with its strengths and weaknesses and the debate about the relative merits of these methods is not closed. For some species groups, the choice of a method is irrelevant: all methods reconstruct the same tree phylogenetics. This is obviously an optimistic case, rarely met in practice. The method of maximum likelihood is much slower than its competitors but provides a framework for a natural additional test of hypotheses so as to quantify the variability of the tree view.

## 11.3    Validation of Phylogenetic Trees

As in most statistical inference procedures, the estimated tree depends on the data: the same estimation procedure applied to different sets of data will give different trees. It is essential to quantify this variability, for example, by proving that the tree view is not very different from the true tree. The standard way is to prove a limit theorem on the tree view, usually a theorem based on asymptotic normality of the distance between the estimated and the true tree.

But a phylogenetic tree is an unusual setting: it is composed of a discrete topology (tree shape) and continuous length of branches, which depend on the topology of the tree. Phylogenetic tree spacing has a more complex structure (Billera et al. [25]) that makes tools used to establish convergence limit theorems ineffective.

Without limit theorems, the variability of the estimator is generally quantified using resampling techniques, such as bootstrap or jackknife that mimic the independent samples (Felsenstein 1973 [118]). See Figure 11.2 for an example.

It is necessary to validate the robustness of the estimated tree. The alignment errors and sequencing can indeed generate small changes in the dataset. What is the influence of these small changes on the estimated tree? If their influence is small, the tree estimate is robust to sequencing errors and alignment: it is legitimate to use in subsequent analyses. In the opposite case, the tree is not robust: the analyses based on this tree are unreliable. Again, bootstrap methods and the jackknife can quantify the robustness of the tree.

In the case of not-robust trees, it is interesting to identify erroneous data for correcting or deleting part of the dataset. The sequencing errors and alignments tend to create data, very different from the rest of the dataset and therefore unexpected. The framework of maximum likelihood is not used only to quantify the variability of the tree estimated but also the exceptional nature or not of a given site. It is therefore particularly suitable for the detection of erroneous data ([12, 222]).

### 11.3.1   Influential Sites

An inference method should be robust to small violations of the model. It should also be resistant to outliers. Indeed, if the method is overly sensitive to the data, changing as few as a handful of sites can dramatically modify the inferred tree. The significance of such a tree should be seriously questioned. There are several ways to construct outlier resistant methods.

Since we want to characterize the influence of each site on the likelihood, it is crucial to study them one at a time. Let $T$ be the tree that maximizes the likelihood of the whole dataset and $T^{(h)}$ be the tree that maximizes the likelihood of the jackknife sample obtained when removing site $h$ from the original dataset. By comparing $T$ to each $T^{(h)}$, we study the impact of each site on $T$ and can relate the stability or lack of a stability of a clade to a particular site or set of sites. We also define the outlier sites as those whose influence values are the greatest. Outlier sites may arise from biologically well-known characteristics that result in evolution schemes not taken into account by the evolution model, such as the nature of the mutation of guanine-cytosine (GC)-content for a given nucleotide dataset. Taking a further step toward robustness, we order the sites in the original dataset from the strongest outlier to the weakest outlier and remove them one at a time, starting with the strongest outlier. Doing so, we obtain a sequence of samples, each one shorter than the previous one by exactly one nucleotide, from which the corresponding sequence of trees is inferred. Assuming that major causes of disruption and thus instability disappear along with the strongest outlier, we expect a stable tree to arise from this sequence.

The influence function of each site was computed from an alignment of the small subunit (SSU) ribosomal (rRNA) gene (1026 nucleotides) over 157 terminals (i.e., 157 rows), all fungi belonging to the phyla Chytridiomycota, Zygomycota, Glomeromycota plus one outgroup to root the tree, Corallochytrium limacisporum, a putative choanoflagellate. This alignment was chosen to satisfy different criteria: (i) enough variation accumulated to clearly resolve the phylogenetic topology, (ii) a very low number of detectable homoplasic events, (iii) a strong monophyletic group (i.e., Glomeromycota), (iv) a highly polyphyletic group (i.e., Zygomycota), and (v) one group with uncertainties about phylogenetic affinities (i.e., Chytridiomycota).

When removing a site, between 11 and 32 internal nodes of the ML (maximum likelihood) tree were affected (see Figure 11.2). An average of 15 nodes is affected by removing only one site. These nodes were related to terminals with high homology within unresolved clades, i.e., not well supported by the ML tree. Some areas contained the strongest outliers, which were not uniformly distributed along the sequence.

For example, the most infuential site (i.e., strongest outlier) corresponded to a highly variable site. To visualize the position of this particular site we computed the most probable RNA secondary structure (RNA folding) using a method based on thermodynamic principles. From two different sequences selected randomly, and using different temperatures and different salinities, we always found that the strongest outlier is on a small loop (5 nucleotides) carried by a conserved hairpin (figure not shown, available on request).

In order to achieve a more robust tree, we removed the strongest outliers from the analysis. If the outliers indeed disrupt the inferred topology, we expect that, after discarding enough of them, the inferred tree will not be oversensitive to the sample anymore, i.e., removing or adding one site from the analysis will not drastically change it. In order to test this belief, we classified the outliers according to their influence values, from the most negative to the least negative. We then deleted the $i$ strongest outliers (for values of $i$ ranging from 1 to 325) and inferred the ML-GTR (maximum likelihood, general time reversible) tree. Using the Penny-Hendy distance, we quantified the topological similarity of these 325 trees with each other and with the guide tree $T$. Penny-Hendy distance between two phylogenies

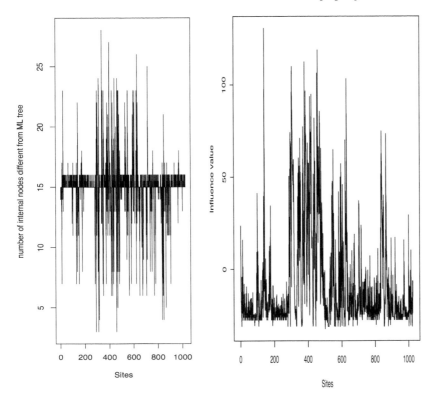

**FIGURE 11.2**
(Left) Number of internal nodes different from the ML-GTR guide tree (all data included) when removing one site only from the dataset. (Right) influence values when removing each of the single sites (i.e., one column only) from the dataset (1026 columns in total).

calculates the minimal composition of elementary mutations that converts the first tree into the second one. From the dataset, we demonstrated that there were two stable trees. Removing any number between 2 and 44 of the strongest outliers led to almost the same tree. This is illustrated by the very small Penny-Hendy distance between these topologies. After removing the 46 strongest outliers an additional stable topology was found but the tree-to-tree distance increased quickly after removing 50 sites leading to unstable phylogenies (see Figure 11.3).

## 11.3.2 Influential Taxons

Sites in the alignment play an important role in providing the information necessary to unravel the (evolutionary) structure of the taxa included in the data matrix. However, taxa also play a full-fledged part in the inference and adequate taxon sampling is all but superfluous. Small changes in the taxon sampling can have dramatic changes on the topology. Like with sites, a tree overly sensitive to taxon sampling should be seriously questioned. Again, constructing phylogenies resistant to rogue species requires preliminary detection of influent species.

The most striking illustration is perhaps the decision to include or not include an outgroup. The outgroup is a taxon or set of taxa distant enough from the rest of the taxa (the ingroup) to ensure than the most recent common ancestor (MRCA) of the ingroup is more

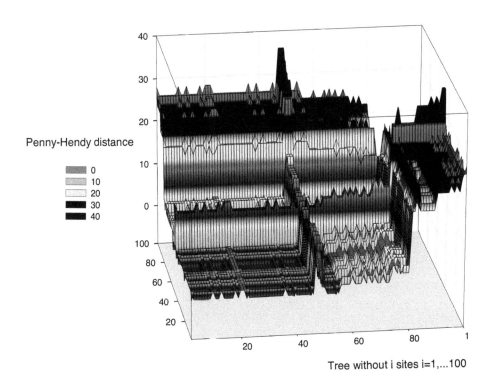

**FIGURE 11.3**
Penny-Hendy tree-to-tree distances. On the $x$ and $y$-axes are plotted the trees inferred after removing the $i$ strongest outliers ($i = 1, \ldots, 100$). The ML-GTR guide tree (all data included, i.e., $i = 0$) is not included in this figure.

recent than the MRCA of all the taxa, and therefore to root the phylogeny. It is known that the inclusion of the outgroup in the analysis may disrupt the ingroup phylogeny. Another bias has been well documented: when two nonadjacent taxa share many character states along long branches because of convergence, some inference methods often interpret such a similarity as homology. The resulting inferred tree displays the two taxa as sister taxa, attributing the shared changes to a branch joining them. This effect is termed Long Branch Attraction (LBA) and causes some methods, most notably parsimony, to be inconsistent, meaning they converge to an incorrect tree as the number of characters used in the dataset increases. But extreme sensitivity of the estimates to taxon inclusion has no reason to be limited to outgroup or taxa at the end of long branches. Some authors even recommend that "analysis of sensitivity to taxon inclusion [...] should be a part of any careful and thorough phylogenetic analysis." And there is still more. In addition to extreme sensitivity, phylogenetic analysis of few taxa (but each represented by many characters) can be subject to strong systematic biases, which in turn produce high measures of repeatability (such as bootstrap proportions) in support of incorrect or misleading phylogenetic results.

A natural way to measure the influence of a species is via jackknifing of species. Unlike sites, jackknifing of species is a peculiar statistical problem as species are not independent. Therefore, the statistical framework for jackknifing species is far more complex than its equivalent for jackknifing sites.

We introduce Taxon Influence Index (TII), a resampling method focused on the detection of highly influential taxa. TII quantifies the influence of a taxon on the phylogeny estimate by dropping it from the analysis and quantifying the resulting modifications of the inferred phylogeny. We also introduce the stability of a branch with respect to taxon sampling, defined as the number of taxa that can be dropped without altering the branch.

We examined an empirical dataset taken from Kitazoe et al. (2007) (Kitazoe et al. 2007 [183]) consisting of mitochondrial protein sequences (3658 amino acid sites in total) from 68 mammals species, belonging to Laurasiathera, Supraprimates, Xenartha, and Afrotheria plus seven outgroup taxa, belonging to Monotremata and Marsupialia. The gaps are excluded and the sequences are not realigned when removing a taxon.

As pointed out by Kitazoe et al., these data present relatively long sequences, good taxon sampling, and very little missing data. Another advantage of mammals is that their phylogeny has been intensively studied and that many problems and hard-to-resolve clades have been identified. Of particular interest is the position of the guinea pig (Cavia porcellus) in the order Rodentia, which has long been a heated issue among molecular phylogeneticists.

While TII is amenable to phylogenetic inference method we restricted the analysis to maximum likelihood (ML) for the sake of brevity.

As we use ML, the inferred tree $T_i$ should have a higher likelihood than $T_i^*$. After all, $T_i$ is inferred to maximize the likelihood over $\mathbf{X}^{(i)}$ whereas $T_i^*$ maximizes the likelihood over $\mathbf{X}$ before being pruned. Although we expect the likelihood values of $T_i$ and $T_i^*$ to be close and even to correlate quite well as only a fraction of the taxa strongly affect the inference, they should systematically be higher for $T_i$ than for $T_i^*$. Results from our analyses (data not shown) confirm this.

Phylogenetic trees were inferred using PhyML 2.4.4 (Guindon & Gascuel 2003 [150]). We used the mtMam+I+$\Gamma$4 model, selected by ProtTest 1.4 as the best model no matter what the criterion [Akaike information criterion (AIC), Bayesion information criterion (BIC), etc.] was. The mtMam empirical rate matrix is the one used in the best four models (mtMAM and any combination of +I and +$\Gamma$4), followed by mtREV in the same four models. The hill-climbing search was initiated at the BIONJ tree of the alignment, the default starting tree for PhyML. We used PhyML to optimize the branch lengths, the $\alpha$ parameter of the $\Gamma$ shape and the proportion of invariable sites (command line: `phyml alignment 1 s 68 0 mtMAM e 4 e BIONJ y y`). Thanks to the moderate size of the dataset (68 species, 3658

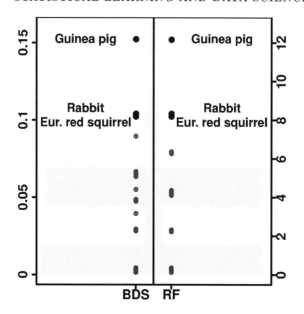

**FIGURE 11.4**
Dot plot and histogram of TII (taxon influence index) values for BSD (left) and RF (right) distance. Taxa with TII higher $\geq 0.75$ (BSD) or $\geq 8$ (RF) are labeled with their name. Though many taxa have the same TII, and thus should have exactly the same position in the dot plot, the corresponding superimposed points are slightly shifted for better legibility.

AA), each of the 69 trees (1 for the whole dataset and 1 for pruning each of the 68 species) was inferred in $\sim$ 45 minutes CPU time (on a 1.66-GHz Intel Core Duo PC). We also performed 200 replicate ML bootstrap analyses for the whole data set (68 species) in a bit more that 3 CPU days.

Schmidt-Lanterman incisures (SLI) values of the species are plotted in Figure 11.4. We note that guinea pig has the highest TII (12), confirming previous findings of guinea pig being a rogue taxon. The result is robust to model choice (with or without Rate Across Sites (RAS), and with mtREV instead of mtMAM), with guinea pig TII always being the highest, between 12 and 14. The comparison of the pruned and inferred tree (not shown) for the guinea pig reveals that removing as little as one species can affect the topology even in remote places; removing the guinea pig disrupts the clades of the insectivores and modifies the position of the northern tree shrew (*Tupaia belangeri*), 6 branches away from it.

Using a cutoff value of 8, which represents two standard deviations from the mean, three species are identified as influential (marked in bold and annotated in Figure 11.4) and concentrated among *Glires*: guinea pig, European red squirrel (*Sciurus vulgaris*), and rabbit (*Oryctolagus cuniculus*). No matter what distance is used [Robinson-Foulds (RF) or Branch score distance (BSD)], the same species stand out as influential (Figure 11.4) and the TII-induced order is conserved; only 4 of the remaining 65 species change rank when changing the distance. But the number of influential species is highly dependent on the model: it varies from 4 for the mtMam+I+$\Gamma$ to 10 $\sim$ 12 for mtMam+$\Gamma$ and mtREV+$\Gamma$. Fortunately, there is important overlap; for example, the 4 species influential under mtMam+I+$\Gamma$ are part of the set of species influential under mtMam+$\Gamma$. Conversely, 20 species (again varying with the model from 7 in mtREV/mtMAM+$\Gamma$ to 20 in mtMam+I+$\Gamma$), in bold in Figure 11.4, are extremely stable in the sense that their removal does not disturb the topology at all.

Remarkably, the stable species are well distributed over the tree and either part of a clade of two species or at the end of a long branch.

With the exception of influential species and extremely stable species, most of the TII values are 4. This means that most inferred trees are slightly different from the corresponding pruned trees, with a difference of only two branches. We use the stability scores to check whether these differences are well distributed over the whole topology $T^*$ or concentrated on a limited number of branches. The results are shown in Figure 11.5 (inset). Interestingly there is no correlation between branch stability (BS) scores and branch lengths (data not shown) even when restricting the analysis to the branches with BS $< 100\%$.

Two branches with very low BS scores belong to the *Afrotheria* (Figure 11.5), indicating a poorly resolved clade. Indeed, even if a species is only weakly connected to the *Afrotheria*, removing it from the analysis often changes the inner configuration of the *Afrotheria* clade. These branches have very low bootstrap values (from the root to the leaves 45%, 54%). A detailed comparison between BS scores and bootstrap values is informative about their similarities and differences. First, bootstrap is more conservative than BS: all branches with 100% bootstrap values also have 100% BS, but some branches (20) with 100% BS do not have 100% bootstrap values (marked in light gray in Figure 11.5). Second, for the 9 branches whose both BS and bootstrap values are lower than 100% (marked in dark gray in Figure 11.5), the correlation is very low (0.25). Except for the two branches aforementioned, the bootstrap values are much smaller than their BS equivalent: they vary between 11% and 75% whereas all BS scores are over 92%.

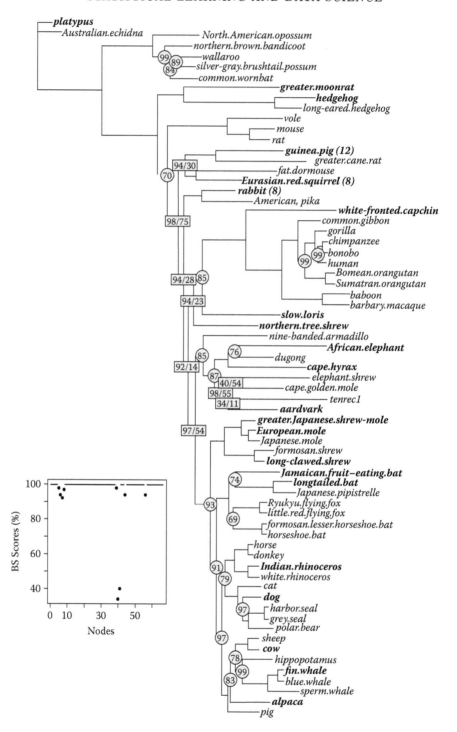

**FIGURE 11.5**

Placental mammals phylogeny with BS scores and bootstrap values. 100% bootstrap and BS values are omitted. Light gray branches have 100% BS scores but <100% bootstrap (circles), whereas dark gray branches have <100% BS and bootstrap scores (resp. left and right in the rectangle). Influential taxa (TII>7) are in bold and annotated by their TII, stable taxa are in bold. Inset: BS scores (in %) of internal branches.

# 12

## Bayesian Analysis of Structural Equation Models Using Parameter Expansion

Séverine Demeyer, Jean-Louis Foulley, Nicolas Fischer, and Gilbert Saporta

## CONTENTS

## 12.1 Introduction

### 12.1.1 From Latent Variables to Structural Equation Models (SEM)

This chapter relies on the ambivalent nature of latent variables. Their unobserved nature makes them either auxiliary variables used as computational tricks or latent concepts associated with observed variables.

This chapter combines the power of these two aspects of latent variables.

Precisely latent auxiliary variables have proven to be effective computational tools when applied to the expectation maximization (EM) algorithm, see Dempster, Laird, Rubin (Dempster et al. 1977 [87]), and to data augmentation, see Tanner and Wong (Tanner

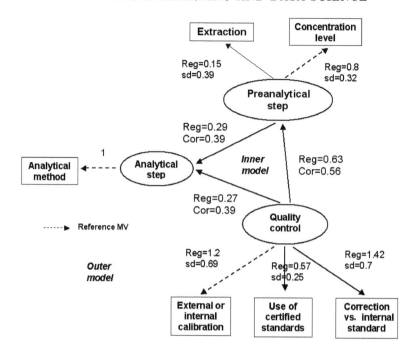

**FIGURE 12.1**
Estimates of SEM used to model expert knowledge in the water pollutants field. *Reg* is
the estimated regression coefficient, *sd* is the associated standard deviation and *cor* is the
correlation coefficient. The dotted arrows represent reference variables defining the sign of
latent variables.

& Wong 1987 [290]), and even more effective when implemented in EM with parameter
expansion (PX), see Liu, Rubin, Wu (Liu et al. 1998 [211]) and in MCMC with PX-Gibbs
sampling, see Liu and Wu (Liu & Wu 1999 [212]) and van Dyk (van Dyk & Meng 2001 [296]).

On the other side, using meaningful latent variables is very popular in applied domains
like marketing, psychology, sociology, and education where interest lies in quantifying unob-
servable characteristics or aptitudes of individuals like satisfaction, self esteem, alienation,
and aptitude at school from studies.

Such latent constructs may also involve several meaningful latent variables associated
with observed variables, thus focusing the interest on the relationships between these latent
variables. The causal relationships between these latent variables are supposed to reflect
the structure in the observed variables. Hence, the terminology structural equation models
(SEM).

In other words SEM are multivariate latent variable models used to represent causal
latent structures in the data. The observed (manifest) variables are associated with latent
variables in the outer (measurement) model and causality links are assumed between latent
variables in the inner (structural) model, see Figure 12.1.

## 12.1.2  Motivation of a Bayesian Approach to SEM

The Bayesian approach of this chapter has been motivated by our own practice of SEM, to
take advantage of the information conveyed in structural latent variables, processed outside

the SEM. In that respect we are especially interested in the prediction of the structural latent variables.

Bayesian estimation of SEM meets this requirement, providing draws from the joint posterior distribution of latent variables that are directly reusable outside SEM.

### 12.1.3    Motivation of a Parameter Expansion (PX) Framework

In this chapter PX is used to overcome identifiability issues due to the unobserved nature of latent variables. Identifiability issues of SEM are overcome by setting a scale for the latent variables. This issue has been addressed by Skrondal and Rabe-Hesketh (Skrondal & Rabe-Hesketh 2004 [285]) who propose to either scale latent variables in terms of a chosen manifest variable in each block (*anchoring*) or standardize latent variables (*scaling*).

A Bayesian approach to SEM has already been proposed by Lee (Lee 2007 [201]) under *anchoring*, making however the imputation of structural latent variables somewhat tedious.

Using parameter expansion instead allows to easily sample the covariance matrix of latent variables as a correlation matrix (see sections 12.3.1 and 12.3.3) thus overcoming identifiability issues.

## 12.2    Specification of SEM for Mixed Observed Variables

### 12.2.1    Measurement (Outer) Model

Let $\mathbf{Y_i}$ be the row vector of mixed continuous, binary and ordered categorical observed outcomes for individual $i$ on the $p$ manifest variables, divided into $q$ disjoint blocks indexed by $k = 1 \ldots q$ and $n_k$ the number of observed variables within block $k$. Each block is assumed to reflect a unidimensional concept, summarized into a unique continuous latent variable. Let $\mathbf{Z_i}$ be the row vector of $q$ continuous latent variables for individual $i$.

SEM with mixed observed variables is defined within the framework of generalized linear models where binary and ordered categorical observed variables (indexed by $j = 1 \ldots n_k$) are modeled as latent responses following Albert and Chib (Albert & Chib 1993 [4]), using probit link functions and threshold values.

Let $\mathbf{Y_i^*} = \left\{ Y_{ikj}^*, \, k = 1 \ldots q, \, j = 1 \ldots n_k \right\}$ be the row vector of latent responses defined in expressions 12.3 and 12.4.

The measurement model relates each latent response vector to its associated structural latent variable in a reflexive model (because each observed variable reflects its latent variable) where conditional independence of observed variables is assumed given latent variables.

Using matricial notations the outer model is written for individual $i$ as

$$\mathbf{Y_i^*} = \boldsymbol{\mu} + \mathbf{Z_i}\boldsymbol{\theta} + \mathbf{E_i}, \, 1 \leq i \leq n \tag{12.1}$$

where $\mathbf{E_i}$ is the measurement error term distributed $\mathbf{E_i} \sim \mathcal{N}\left(0, \boldsymbol{\Sigma}_\varepsilon\right)$ with $\boldsymbol{\Sigma}_\varepsilon$ diagonal and $\boldsymbol{\theta}$ is the $q \times p$ matrix of regression coefficients.

To illustrate the formulas, with $q = 3$, $p = 6$, $n_1 = 2$, $n_2 = 2$ and $n_3 = 3$ (see the graphical model section 12.4.5) $\boldsymbol{\theta}$ is the matrix

$$\boldsymbol{\theta} = \begin{pmatrix} \theta_{11} & 0 & 0 & 0 & 0 & 0 \\ 0 & \theta_{22} & \theta_{23} & 0 & 0 & 0 \\ 0 & 0 & 0 & \theta_{34} & \theta_{35} & \theta_{36} \end{pmatrix} \tag{12.2}$$

If $\mathbf{Y}_{kj}$ is continuous then it coincides with its quantified version $\mathbf{Y}^*_{kj}$.

If $\mathbf{Y}_{kj}$ is binary or ordered categorical then $\mathbf{Y}^*_{kj}$ is defined in the following univariate probit models.

A probit link for binary variables is used to model the probability of success $p(Y_{ikj} = 1) = \Phi(\mu_{kj} + \theta_{kj} Z_{ik})$.

The univariate probit model for binary outcomes is written

$$
\begin{aligned}
Y_{ikj} &= 1_{\{Y^*_{ikj} \geq 0\}} \\
Y^*_{ikj} &\sim \mathrm{N}(\mu_{kj} + \theta_{kj} Z_{ik}, 1)
\end{aligned}
\tag{12.3}
$$

A probit link for ordered categorical variables is used to model the cumulated probabilities $p(Y_{kj} \leq c) = \Phi(\gamma_{kj,c} + \theta_{kj} Z_k)$.

The univariate probit models for ordered categorical outcomes is written

$$
\begin{aligned}
Y_{ikj} &= c \iff \gamma_{kj,c-1} < Y^*_{ikj} \leq \gamma_{kj,c} \\
Y^*_{ikj} &\sim \mathrm{N}(\theta_{kj} Z_{ik}, 1)
\end{aligned}
\tag{12.4}
$$

where, to ensure identifiability of thresholds, $\gamma_{kj,0} = -\infty$, $\gamma_{kj,1} = 0$ and $\gamma_{kj,n_{kj}} = \infty$ ( $n_{kj}$ the number of categories of question $kj$).

If $\mathbf{Y}^*_{\mathbf{i}}$ and $\mathbf{Z}_{\mathbf{i}}$ were observed, the measurement model (12.1) would reduce to a linear regression model.

## 12.2.2    Structural (Inner) Model: Alternative Modeling

Denoting $\mathbf{H_i}$ the endogenous latent variables and $\mathbf{\Xi_i}$ the exogenous latent variables, the structural equations are simultaneous equations given by

$$
\mathbf{H_i} = \mathbf{H_i}\mathbf{\Pi} + \mathbf{\Xi_i}\mathbf{\Gamma} + \mathbf{\Delta_i}
\tag{12.5}
$$

where $\mathbf{Z_i} = \mathbf{H_i}\mathbf{\Xi_i}$, $\mathbf{\Pi}$ is the $q_1 \times q_1$ matrix of regression coefficients between endogenous latent variables , $\Gamma$ is the $q_2 \times q_1$ matrix of regression coefficients between endogenous and exogenous latent variables. $\mathbf{\Delta_i}$ is the error term distributed $\mathbf{\Delta_i} \sim \mathcal{N}(0, \mathbf{\Sigma_\delta})$ with $\mathbf{\Sigma_\delta}$ diagonal, independent with $\mathbf{\Xi_i}$ and $\mathbf{\Xi_i}$ is distributed $\mathcal{N}(0, \mathbf{\Phi})$.

Since a Bayesian approach works with the joint distribution of latent variables, it is equivalent to work with the correlation matrix of latent variables, under the assumption of multinormality for the conditional distribution of $\mathbf{Z_i}$, so that the inner model considered in this chapter is given by

$$
\mathbf{Z_i} | \mathbf{R_Z} \sim N(\mathbf{0}, \mathbf{R_Z})
\tag{12.6}
$$

with $\mathbf{R_Z}$ a correlation matrix.

In addition $\mathbf{R_Z}^{-1}$ contains the regression parameters of all possible regressions between latent variables.

## 12.3    Bayesian Estimation of SEMs with Mixed Observed Variables

### 12.3.1    Implementation of Parameter Expansion

The implementation of parameter expansion in this chapter mimics the implementation of parameter expansion in PX-EM algorithms as defined in Liu, Rubin, Wu (Liu et al. 1998 [211]) and so differs from the usual implementation for Monte Carlo Markov chain (MCMC) algorithms described in (Liu & Wu 1999 [212]).

Parameter expansion (Liu et al. 1998 [211]) consists of working with unidentified parameters in the complete data model $f(\mathbf{Y}, \mathbf{Z}|\boldsymbol{\theta})$ and indexing expanded latent variables $\mathbf{W}$ and expanded data models $p(\mathbf{Y}, \mathbf{W}|\boldsymbol{\theta}, \boldsymbol{\alpha})$ each corresponding to a value of the expansion parameter $\boldsymbol{\alpha}$, so that the observed likelihood $f(\mathbf{Y}|\boldsymbol{\theta})$ is preserved, that is, satisfies

$$f(\mathbf{Y}|\boldsymbol{\theta}) = \int f(\mathbf{Y}, \mathbf{Z}|\boldsymbol{\theta}) \, \mathrm{d}\mathbf{Z} = \int p(\mathbf{Y}, \mathbf{W}|\boldsymbol{\theta}, \boldsymbol{\alpha}) \, \mathrm{d}\mathbf{W} \qquad (12.7)$$

Usually, the transformation indexed by the expansion parameter is a $C^1$ diffeomorphism (one-to-one mapping).

In this chapter, the variances of structural latent variables $\mathbf{Z} = \mathbf{Z_1}, \ldots, \mathbf{Z_q}$ are expansion parameters. Under scaling constraints (that is in the complete data model), recall that $\mathbf{Z} \sim N(0, \mathbf{R_Z})$ where $\mathbf{R_Z}$ is a correlation matrix. Introducing variance parameters $\alpha_1, \ldots, \alpha_q$ and defining $\boldsymbol{\alpha} = diag(\alpha_1, \ldots, \alpha_q)$, creates expanded latent variables $\mathbf{W} = \boldsymbol{\alpha}^{\frac{1}{2}}\mathbf{Z}$ in the expanded data model indexed by $\boldsymbol{\alpha}$ where $\mathbf{W} \sim N(0, \boldsymbol{\Sigma_W})$ with $\boldsymbol{\Sigma_W} = \boldsymbol{\alpha}^{\frac{1}{2}}\mathbf{R_Z}\boldsymbol{\alpha}^{\frac{1}{2}}$ a covariance matrix.

Identifiability issues are easily overcome in the parameter expansion setting: drawing a correlation matrix in the complete data model only involves sampling a covariance matrix in the expanded data model and applying the reverse transformation to the covariance matrix $\mathbf{R_Z} = \boldsymbol{\alpha}^{-\frac{1}{2}}\boldsymbol{\Sigma_W}\boldsymbol{\alpha}^{-\frac{1}{2}}$.

The same applies with residual variances of latent responses, where the expansion parameter $\alpha$ is the residual variance, see Meza, Jaffrézic, Foulley (Meza et al. 2009 [227]).

Parameter expansion involves computation of conditional posterior distributions of original and expansion parameters, where the expansion parameters are computed in the expanded data model and the original parameters are computed in the complete data model.

### 12.3.2    Imputation of Latent Variables

Latent responses are computed following Albert and Chib (Albert & Chib 1993 [4]) based on models 12.3 and 12.4 for binary and ordered categorical variables, respectively.

For a binary observed variables, $Y_{ikj}^*$ is drawn from

$$Y_{ikj}^*|\mu_{kj}, \theta_{kj}, Z_{ik}, Y_{ikj} \sim \mathrm{NT}(\mu_{kj} + \theta_{kj}Z_{ik}, 1; 0, \infty) \text{ si } Y_{ikj} = 1 \qquad (12.8)$$

$$Y_{ikj}^*|\mu_{kj}, \theta, Z_{ik}, Y_{ikj} \sim \mathrm{NT}(\mu_{kj} + \theta_{kj}Z_{ik}, 1; -\infty, 0)) \text{ si } Y_{ikj} = 0 \qquad (12.9)$$

where $NT(\mu, 1; a, b)$ stands for the normal distribution $N(\mu, 1)$ left truncated at $a$ and right truncated at $b$.

For ordered categorical variables, the latent response $Y_{ikj}^*$ is drawn from

$$Y_{ikj}^*|, \theta, Z_{ik}, Y_{ikj} \sim \mathrm{NT}\left(\theta_{kj}Z_{ik}, 1; \gamma_{kj, Y_{ikj}-1}, \gamma_{kj, Y_{ikj}}\right) \qquad (12.10)$$

Given $\Theta = \{\mu, \theta, \Sigma_\varepsilon, \mathbf{R_Z}\}$ the conditional posterior distribution of latent variables is expressed as

$$[\mathbf{W}_i | \mathbf{Y}_i^*, \Theta] \propto [\mathbf{Y}_i^* | \mathbf{Z_i}, \Theta] \, [\mathbf{Z}_i | \Theta] \tag{12.11}$$

$$\propto [\mathbf{Y}_i^* | \mathbf{Z}_i, \mu, \theta, \Sigma_\varepsilon] \, [\mathbf{Z}_i | \mathbf{R_Z}] \tag{12.12}$$

where $\mathbf{Y}_i^* | \mathbf{Z}_i, \mu, \theta, \Sigma_\varepsilon \sim N(\mu + \theta \mathbf{Z}_i, \Sigma_\varepsilon)$ is the likelihood of individual $i$ computed from the measurement model (12.1) and $\mathbf{Z_i} | \mathbf{R_Z} \sim N(\mathbf{0}, \mathbf{R_Z})$ is the joint distribution of latent variables.

Then it can be easily shown that

$$\mathbf{W}_i | \mathbf{Y}_i^*, \mu, \theta, \Sigma_\varepsilon, \mathbf{R_Z} \sim \mathcal{N}\left(\mathbf{D}\theta\Sigma_\varepsilon^{-1}(\mathbf{Y}_i^* - \mu), \mathbf{D}\right) \tag{12.13}$$

where $\mathbf{D}^{-1} = \theta\Sigma_\varepsilon^{-1}\theta^t + \mathbf{R_Z}^{-1}$.

### 12.3.3    Simulation of the Covariance Matrix of Structural Latent Variables

The covariance matrix $\Sigma_W$ of structural latent variables is computed in the expanded data model under the following conjugate prior distribution

$$\Sigma_W \sim \text{Inverse-Wishart}_{\nu_0}\left((\nu_0 \mathbf{S_0})^{-1}\right) \tag{12.14}$$

where $\nu_0$ is a degree of freedom and $\mathbf{S_0}$ is our prior guess at covariance matrix $\Sigma_W$. A weakly informative prior is given by $\nu_0 = q$ or $q + 1$.

The posterior distribution of $\Sigma_W$ is given by

$$\Sigma_W | \mathbf{W} \sim \text{Inverse-Wishart}_{\nu_0 + n}\left((\nu_0 \mathbf{S_0} + \mathbf{W^t W})^{-1}\right) \tag{12.15}$$

### 12.3.4    Conditional Posterior Distributions in the Measurement Model

**Conditional posterior distributions of regression parameters**

The model relating each latent response to its associated latent variable is a linear regression model, whose conjugate prior distribution is factorized as

$$[\mu_{kj}, \theta_{kj}, \sigma_{kj}^2] = [\mu_{kj}, \theta_{kj} | \sigma_{kj}^2] \, [\sigma_{kj}^2] \tag{12.16}$$

Under the conjugate prior distributions

$$\mu_{kj}, \theta_{kj} | \sigma_{kj}^2 \sim N\left(\begin{pmatrix}\mu_{0kj} \\ \theta_{0kj}\end{pmatrix}, \sigma_{kj}^2 \mathbf{H_0}^{-1}\right) \tag{12.17}$$

$$\sigma_{kj}^2 \sim \text{Inverse-Gamma}\left(\frac{\nu_0}{2}, \frac{\nu_0}{2} s_0^2\right) \tag{12.18}$$

with $\mathbf{H_0}^{-1}$ the $2 \times 2$ prior covariance matrix of $(\mu_{kj}, \theta_{kj})$, and $s_0^2$ our prior guess for the residual variance $\sigma_{kj}^2$. A weakly informative prior is given by $\nu_0 = 1$ or 2.

Computation gives, with $\mathbf{X_k} = (\mathbf{1}, \mathbf{Z_k})$, $\boldsymbol{\beta_{kj}} = (\mu_{kj}, \theta_{kj})$, $\boldsymbol{\beta_{0kj}} = (\mu_{0kj}, \theta_{0kj})$

$$\beta_{kj}|\sigma_{kj}^2, \mathbf{Y}_{kj}^*, \mathbf{Z_k} \quad \sim \quad N\left(\left(\mathbf{X_k^t X_k} + \mathbf{H_0}\right)^{-1}\left(\mathbf{X_k^t Y}_{kj}^* + \mathbf{H_0}\beta_0\right), \sigma_{kj}^2\left(\mathbf{X_k^t X_k} + \mathbf{H_0}\right)^{-1}\right)$$
$$(12.19)$$

$$\sigma_{kj}^2|\mathbf{Y}_{kj}^*, \mathbf{Z_k^*} \sim \text{Inverse-Gamma}\left(\frac{1}{2} + \frac{n}{2}, \frac{1}{2} + \frac{n}{2}\tilde{s}_{kj}^2\right) \qquad (12.20)$$

$$n\tilde{s}_{kj}^2 = \left(\mathbf{Y}_{kj}^* - \mathbf{X_k}\beta_{kj}\right)^t\left(\mathbf{Y}_{kj}^* - \mathbf{X_k}\beta_{kj}\right) + \left(\beta_{kj} - \beta_{0kj}\right)^t H_0\left(\beta_{kj} - \beta_{0kj}\right) \qquad (12.21)$$

**Conditional posterior distributions of thresholds (for categorical observed variables)**

Thresholds are defined in model 12.4 where threshold $\gamma_{kj,c}$ separates modalities $c$ and $c+1$. Assuming flat prior distributions $[\gamma_{kj,c}] \propto 1$, the posterior distribution of threshold $\gamma_{kj,c}$ for $2 \le c \le n_{kj} - 1$ is given by

$$\gamma_{kj,c}|Y_{kj}, Y_{kj}^*, \{\gamma_{kj,c'}, c' \ne c\} \quad \sim \quad \text{Unif}\left(\max\left\{Y_{kj}^* : Y_{kj} = c\right\}, \min\left\{Y_{kj}^* : Y_{kj} = c+1\right\}\right)$$
$$(12.22)$$

which retrieves a stochastic EM estimate of threshold values. An alternative proposition would be to assume, as in Foulley and Jaffrézic (Foulley & Jaffrézic 2010 [135]), that the $\Delta_{kj,c} = \gamma_{kj,c} - \gamma_{kj,c-1}$ are uniformly distributed on the range $[0, \delta]$.

### 12.3.5  PX-Gibbs Sampling

PX-Gibbs algorithm for estimating SEM with mixed outcomes involves two PX schemes yielding a three-step algorithm described as follows, whose steps are similar to the homologous steps implemented in PX-EM.

- **Step 1:** PX implementation in the probit models to generate latent responses matching the constraint of residual variance fixed at unity, given structural latent variables and current values of parameters in the complete data model.

  - Draw latent responses in the expanded data model
    $Y_{ikj}^{*(t+1)} = Y_{ikj}$ if $Y_{ikj}$ is continuous
    $Y_{ikj}^{*(t+1)} \sim f\left(Y_{ikj}^*|\mu_{kj}^{(t)}, \theta_{kj}^{(t)}, \sigma_{kj}^2 = 1, Z_{ik}^{(t)}, Y_{ikj}\right)$ if $Y_{ikj}$ is binary

    $$Y_{ikj}^{*(t+1)} \sim f\left(Y_{ikj}^*|\mu_{kj}^{(t)}, \theta_{kj}^{(t)}, \sigma_{kj}^2 = 1, \gamma_{kj,Y_{ikj}}^{(t)}, \gamma_{kj,Y_{ikj}+1}^{(t)}, Z_{ik}^{(t)}, Y_{ikj}\right)$$
    $$\text{if } Y_{ikj} \text{ is ordered categorical}$$

    where $f$ is a generic notation defined in expressions 12.8, 12.9, and 12.10.
  - Draw the expansion parameters of the probit models

    $$\sigma_{kj}^2 \sim f\left(\sigma_{kj}^2|\mu_{kj}^{(t)}, \theta_{kj}^{(t)}, Y_{ikj}^{*(t+1)}, Z_{ik}^{(t)}\right)$$

    where $f$ is defined in expression 12.20.

– Compute latent responses in the complete data model

$$Y_{ikj}^{*(t+1)} \leftarrow Y_{ikj}^{*(t+1)} / \sqrt{\sigma_{kj}^2}$$

- **Step 2:** PX implementation in the structural model to generate structural latent variables matching the identifiability constraint of the covariance matrix being actually a correlation matrix, given latent responses with unit variance current values of parameters in the complete data model.

  – Draw latent responses in the expanded data model

  $$\mathbf{W}_\mathbf{i}^{(t+1)} \sim f\left(\mathbf{W}_\mathbf{i} | \boldsymbol{\mu}^{(t)}, \boldsymbol{\theta}^{(t)}, \boldsymbol{\Sigma}_\epsilon = \mathbf{I}_\mathbf{q}, \mathbf{Y}^{*(t)}, \mathbf{R}_\mathbf{Z}^{(t)}\right)$$

  according to formula 12.13.

  – Draw the expansion parameters [correlation matrix of structural latent variable (LV)] according to formula 12.15

  $$\boldsymbol{\Sigma}_\mathbf{Z} \sim f\left(\boldsymbol{\Sigma}_\mathbf{Z} | \mathbf{W}\right)$$

  – Compute structural LV in the complete data model

  $$\mathbf{R}_\mathbf{Z}^{(t+1)} = [\mathrm{diag}\left(\boldsymbol{\Sigma}_\mathbf{Z}\right)]^{-\frac{1}{2}} \boldsymbol{\Sigma}_\mathbf{Z} [\mathrm{diag}\left(\boldsymbol{\Sigma}_\mathbf{Z}\right)]^{-\frac{1}{2}}$$
  $$\mathbf{Z}^{(t+1)} = \mathbf{W}^{(t+1)} [\mathrm{diag}\left(\boldsymbol{\Sigma}_\mathbf{Z}\right)]^{-\frac{1}{2}}$$

- **Step 3:** Computation of the outer parameters from their posterior conditional distribution in the complete data model under both constraints.

$$\mu_{kj}^{(t+1)}, \theta_{kj}^{(t+1)} \sim f\left(\mu_{kj}, \theta_{kj} | \mathbf{Y}_{kj}^{*(t)}, \mathbf{Z}_\mathbf{k}^{(t)}\right)$$
$$\gamma_{kj,c}^{(t+1)} \sim f\left(\gamma_{kj,c} | \gamma_{kj,c-1}^{(t)}, \gamma_{kj,c+1}^{(t)}, \mathbf{Y}_{kj}^{*(t)}, \mathbf{Y}_{kj}^{(t)}\right)$$

## 12.4 Application: Modeling Expert Knowledge in Uncertainty Analysis

### 12.4.1 Context of Interlaboratory Comparisons

This section shows an original application of SEM for the first time in uncertainty analysis, in the field of interlaboratory comparisons. Interlaboratory comparisons are external quality controls designed to help laboratories improve their measurement process by measuring the same quantity and comparing their results.

Quality indicators are the consensus value of the comparison, which is the estimated value of the quantity computed from the results of the laboratories, its associated uncertainty and measurement bias, which is for a given laboratory the difference between its result and the consensus value.

If measurement bias were computed with respect to the true value of the quantity then monitoring measurement bias over time would be meaningful and trends could be detected.

Instead, measurement bias intrinsically depends on results through the consensus value. Hence, the need to robustify the consensus value to make it less dependent to observed data.

Current practice to try to overcome this dependence in computing consensus value and its associated uncertainty involve either

- robust algorithms to be less dependent to outliers, called *robust method*,

- computing the consensus value from a subset of expert laboratories, called *expert laboratories*.

The originality of our approach lies in combining advantages of both current approaches into an alternative approach, acting as a postprocessing of robustified results based on a management of expert knowledge using SEM. This new method is called *robustified consensus value*.

### 12.4.2    SEM to Model Prior Distribution of Measurement Bias

Our use of SEM is to model expert knowledge to score laboratories according to the quality of their practice, based on a ranking of categories for each observed variable from the worst to the best practice. Latent variables can be interpreted as components of the overall quality of laboratories and used as prior information on measurement bias with SEM actually modeling the structure of bias.

This broader framework is still Bayesian in that latent variables represent preexisting information independent of measurement results used to update knowledge on all the indicators of the comparison: measurement bias, the consensus value, and its associated uncertainty.

### 12.4.3    Combining SEM with Measurement Results

Latent variables are transformed into weights $w_i$ reflecting the quality of practice to combine with measurement results. Among others a logistic transform can be applied to the sum $s_i$ of latent variables

$$w_i = \frac{\exp s_i}{1 + \exp s_i} \tag{12.23}$$

A standardized robust algorithm from standard NF ISO 13528 (ISO/TC69 2005 [173]) is first applied to measurement results to treat outliers. Raw results are thus transformed into a winsorized sample. This step is not inconsistent with the approach in that a good laboratory will have a high weight even if it is an outlier with respect to the normal distribution. Besides, standards strongly recommend to treat outliers.

At each iteration of the PX-Gibbs algorithm used to estimate SEM, a weighted mean of the winsorized measurement results $x_i^R$ and its variance are computed from the latent variables through the weights

$$x_p^{(t)} = \sum w_i^{(t)} x_i^R \tag{12.24}$$

$$u^2\left(x_p^{(t)}\right) = \sum w_i^{2\,(t)} V\left(x_i^R\right) \tag{12.25}$$

where $V\left(x_i^R\right)$ is the winsorized variance of the sample.

According to the ergodicity theorem, MCMC draws consequently yield full posterior distributions of the weighted mean and its variance.

### 12.4.4    Robustifying the Consensus Value

The consensus value is modeled in a hierarchical model whose first level is normal centered on the weighted mean with variance being the variance of the weighted mean. Two other levels represent the sampling variabilities of the weighted mean and its variance from the Markov Chains released when estimating SEM.

The marginal posterior distribution of the consensus value is computed from Monte Carlo draws according to the following hierarchical model

$$x_c \sim N\left(\mu_w, \sigma_w^2\right) \tag{12.26}$$

$$\mu_w \sim N\left(\mu_{w_0}, \sigma_{w_0}^2\right) \tag{12.27}$$

$$\sigma_w^2 \sim \text{Inverse-Gamma}\left(\frac{\alpha_0}{2}, \frac{\alpha_0}{2}S_0\right) \tag{12.28}$$

where $\mu_{w_0}$ and $\sigma_{w_0}^2$ are identified from the posterior distribution of the weighted mean, and $\alpha_0$ and $S_0$ are identified from the posterior distribution of the variance of the weighted mean.

Since the first level integrates out the variance parameter, the marginal posterior distribution of the consensus value is a Student distribution.

### 12.4.5    Results

The method was applied to the measurement of concentrations of water pollutants.

This analysis was performed on a small number of laboratories (18) regularly involved in this comparison and willing to take part in this study. The auxiliary information was investigated by a questionnaire designed by selected experts from universities and environmental laboratories.

The SEM resulting from expert processing is represented in Figure 12.1 where posterior distributions have been computed under the following prior distributions $\mu_{kj}, \theta_{kj}|\sigma_{kj}^2 \sim$

$N\left(\begin{pmatrix} 0 \\ 0.5 \end{pmatrix}, \sigma_{kj}^2 I_2\right)$, $\upsilon_{kj}^2 \sim \text{Inverse-Gamma}\left(\frac{1}{2}, \frac{1}{2}\right)$, $\Sigma_W \sim \text{Inverse-Wishart}_3\left((I_3))^{-1}\right)$ with $I_2$ and $I_3$ the $2 \times 2$ and $3 \times 3$ identity matrices respectively.

Results and robustified results are given in Table 12.1.

**TABLE 12.1**

Results ($x_i$) and robustified results ($x_i^R$).

| $x_i$ | 31.0 | 50.0 | 32.6 | 57.5 | 50.0 | 36.0 | 40.0 | 20.0 |
|---|---|---|---|---|---|---|---|---|
| $x_i^R$ | 31.0 | 40.6 | 32.6 | 40.6 | 40.6 | 36.0 | 40.0 | 27.1 |

| 36.0 | 28.1 | 34.5 | 42.4 | 25.0 | 33.3 | 32.0 | 39.0 | 30.0 | 31.0 |
|---|---|---|---|---|---|---|---|---|---|
| 36.0 | 28.1 | 34.5 | 40.6 | 27.1 | 33.3 | 32.0 | 39.0 | 30.0 | 31.0 |

The results from the three competing methods are compared with the reference value, called *reference* and its uncertainty provided by Laboraoire National de Métrologie et d'Essais, National Metrology and Testing Laboratory (LNE) in terms of *consensus value*, the *associated uncertainty*, and the 95% *confidence interval* and summarized in Table 12.2.

**Interpretation of results:**

Results for the three methods used to compute the consensus value show consistency between them (intervals overlap) and with the reference value because the four central estimates belong to all the confidence intervals.

**TABLE 12.2**
Consensus value, associated uncertainty, and confidence intervals.

| Method | Cons. Val. | As. Uncert. | 95 % Conf. Int. |
|---|---|---|---|
| Reference | 33 | 1.72 | [29.55, 36.45] |
| Robust method | 34.46 | 1.64 | [31.18, 37.75] |
| Expert labs | 34 | 1.6 | [30.8, 37.2] |
| Robustified CV | 33.82 | 1.62 | [30.57, 37.06] |

Due to the small number of laboratories, estimates of SEM are of poor quality, with relatively high standard deviations, so that this application cannot be used in the general purpose of testing relationships between *preanalytical step*, *analytical step*, and *quality control*.

The implementation of the methodology seems all the same very promising and points out the benefits of additional information to improve existing methods, among them a larger number of laboratories, and results with uncertainties, necessary to quantify sources of measurement bias.

## 12.5     Conclusion and Perspectives

This work applied to uncertainty analysis provides practitioners with a powerful and flexible statistical tool based on Structural Equation Modeling of expert knowledge on measurement bias to improve the treatment of interlaboratory comparison data.

The new method relies on current robust standardized methods to propose a fully Bayesian modeling of interlaboratory comparisons data involving a Bayesian estimation of SEM.

The complete Bayesian framework easily handles missing or censored data as well as a hierarchical structure in the results (e.g., measurements by country) and provides a rigorous framework for model comparison and validation.

The benefits of such a statistical approach are long term and the approach should accompany laboratories in the process of improving their measurements, along with the development of new reference methods by National Metrology Institutes.

## Acknowledgments

The authors particularly thank Eric Parent from AgroParisTech for his involvement from the beginning of this project and his useful comments.

The research within this EURAMET joint research project receives funding from the European Community's Seventh Framework Programme, ERA-NET Plus, under Grant Agreement No. 217257.

# Part III

# Complex Data

# 13

## Clustering Trajectories of a Three-Way Longitudinal Dataset

Mireille Gettler Summa, Bernard Goldfarb, and Maurizio Vichi

## CONTENTS

## 13.1 Introduction

In recent years, growing attention has been paid to the study of multivariate-multioccasion phenomena analyzed through a set $\mathbf{X}$ of $IJT$ values corresponding to $J$ variables, observed on a set of $I$ units, on $T$ different occasions (different times, places, etc.).

The three-way array $\mathbf{X}$ is organized according to three modes: units, variables, and occasions. The most widely collected three-way array is given when, together with units and variables, different time occasions are considered. The temporal repeated observation of the units allows us to evaluate the dynamics of the phenomenon differently from the classical case of a multivariate or cross-sectional (two-way) dataset. There are several major advantages, over conventional cross-sectional or univariate time-series datasets, when we use three-way longitudinal data: the researcher has a large amount of data to describe the phenomenon increasing the degrees of freedom and reducing co-linearity among explanatory variables. This allows us to make inferences about the dynamics of change from cross-sectional evidence.

The three-way longitudinal data may be the result of the following type of observation:

- Repeated recurring surveys, with no overlap of units, e.g., a survey organization repeats a survey on a defined topic, generally with regular time intervals. No overlaps of the sample units are required at different times. Examples of these surveys are given by the repeated analyses made by all Central Bureaux of Statistics.

- Repeated surveys with partial overlap of units. Also these surveys are repeated at regular intervals. The survey design includes rotating units to allow variance reduction, i.e., units are included in the analysis a number of times, and then rotated out of the survey outcome.

- Longitudinal surveys, with no rotation of units. A set of units is followed over time with a survey designed with this aim. These are called panel data and in the current work we will refer to this type of observation.

In addition let us now suppose that units are heterogeneous, i.e., the population, from which the data are observed at time $t$, is composed of $G$ homogeneous disjoint subpopulations. Panel data are usually from a small number of observations over time (short time series) on a usually large number of cross-sectional units like individuals, households, firms, or governments, and frequently characterize economic, demographic, and social phenomena.

The chapter is organized as follows. Section 13.2 briefly lists the notation used; while, Section 13.3 describes three features of a trajectory: trend, velocity, and acceleration. Section 13.4 describes dissimilarities between trajectories, while Section 13.5 illustrates the model used for clustering and the algorithm proposed. Section 13.6 is devoted to the application on the lung cancer data.

## 13.2   Notation

For the convenience of the reader the notation and terminology used is listed here.

- $I, J, T$
  number of units, variables and occasions, respectively.

- $G, Q$
  number of classes, components for variables, respectively.

- $C_1, C_2, \ldots, C_G$
  $G$ clusters of units.

- $\mathbf{X} = [x_{ijt}]$
  $(I \times J \times T)$ three-way data array; where $x_{ijt}$ is the value of the $j^{(th)}$ variable observed on the $i^{(th)}$ object at the $t^{(th)}$ time. On each occasion, the variables are supposed commensurate, if this is not the case the data are supposed standardized.

- $\mathbf{X}_{I,JT}$
  $(I \times JT)$ matrix $\mathbf{X}_{..1}, \mathbf{X}_{..2,...}, \mathbf{X}_{..T}$, i.e., the matricized version of $\mathbf{X}$ with frontal slabs $\mathbf{X}_{..t} = [x_{ijt}]_{IxJ}$ next to each other. It is column standardized.

- $\mathbf{E} = [e_{ijt}]$
  $(I \times J \times T)$ three-way arrays of residuals terms.

- $\mathbf{E}_{I,JT}$
  $(I \times JK)$ matrix $[\mathbf{E}_{..1}, \mathbf{E}_{..2,...}, \mathbf{E}_{..k}]$, i.e., the matricized version of $\mathbf{E}$ with frontal $\mathbf{E}_{..k} = [e_{ijt}]_{IxJ}$ slabs next to each other.

- $\mathbf{U} = [u_{ig}]$

  $(I \times G)$ membership function matrix defining a partition of units, into $G$ classes, where $u_{ig} = 1$ if the $i^{(th)}$ object belongs to class g,

  $u_{ig} = 0$ otherwise. Matrix U is constrained to have only one nonzero element per row.

- $I_g$ cardinality of cluster

  $C_g$, i.e., $I_g = |C_g| = \sum_{i=1}^{I} u_{ig}$ .

- $\overline{\mathbf{X}} = [\overline{x}_{gjt}]$

  $(G \times J \times T))$ three-way centroid array, where $\overline{x}_{gjt}$ is the centroid value of the $j^{(th)}$ variable obtained on the $g^{(th)}$ cluster at the $t^{(th)}$ occasion.

- $\overline{\mathbf{X}}_{G,JT}$

  $(G \times JT)$ centroids matrix, i.e., matricized version of the centroid array $\overline{\mathbf{X}}$ , with frontal slabs $\overline{\mathbf{X}}_{..k} = [\overline{x}_{gjk}]_{G \times J}$ next to each other.

- $\overline{\mathbf{x}}_i, \overline{\mathbf{u}}_i, \overline{\mathbf{e}}_i$

  column vectors representing the $i^{(th)}$ row of $\mathbf{X}$, $\mathbf{U}$, and $\mathbf{E}$, respectively.

- $\overline{\mathbf{x}}_g$

  $g^{(th)}$ row of $\overline{\mathbf{X}}$ , specifying the centroid vector of the $g^{(th)}$ class of the partition of the $I$ objects.

This chapter deals with the problem of partitioning trajectories of a three-way longitudinal data set into classes of homogeneous trajectories.

## 13.3   Trajectories

A time trajectory describes a nonlinear curve in the J+1 dimensional space that has several characteristics; specifically we consider: trend, velocity, and acceleration [101]. For each object $i$, $\mathbf{X}_{i..} \equiv \{x_{ijt} : j = 1, ..., J; t = 1, ..., T\}$ describes a time trajectory of the $i^{(th)}$ object according to the $J$ examined variables. The trajectory $\mathbf{X}_{i..}$ is geometrically represented by $T$-1 segments connecting $T$ points $\mathbf{X}_{i.t.}$ of $\Re^{JT+1}$ (Figure 13.1).

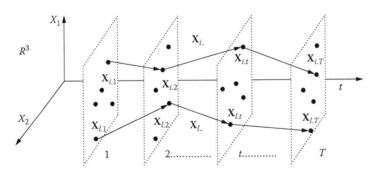

**FIGURE 13.1**

Two time trajectories in $\Re^3$.

Trend is the basic characteristic of a trajectory indicating the tendency of the $J$-variate objects along different time points. Velocity and acceleration are two trajectories' characteristics strongly describing changes of the shape of trajectories. For example in $\Re^2$, velocity of each segment of the trajectory is the slope of the straight line passing through it: if velocity is negative (positive) the slope will be negative (positive) and the angle made by each segment of the trajectory with the positive direction of the $t$-axis will be obtuse (acute). Geometrically, acceleration of each pair of segments of trajectory represents their convexity or concavity. If acceleration is positive (negative) the trajectory of the two segments is convex (concave).

For each time trajectory $\mathbf{X}_{i..}$, the velocity of evolution of an object $i$ in the interval from $t$ to $t+1$ denoted $s_{t,t+1}$ is for the $j^{(th)}$ variable

$$v_{ijt,t+1} = \frac{x_{ijt+1} - x_{ijt}}{s_{t,t+1}}$$

In particular: $v_{ijt,t+1} \gtreqless 0$ $(v_{ijt,t+1} \lesseqgtr 0)$ if object $i$, for the $j^{(th)}$ variable, presents an increasing (decreasing) rate of change of its position in the time interval from $t$ to $t+1$.

$v_{ijt,t+1} = 0$ if the object $i$ for the $j^{(th)}$ variable does not change position from $t$ to $t+1$. Acceleration measures the variation of velocity of $\mathbf{X}_{i..}$ in a fixed time interval. For each time trajectory $\mathbf{X}_{i..}$, the acceleration of an object $i$ in the interval from $t$ to $t+2$ denoted $s_{t,t+2}$ is for the $j^{(th)}$ variable

$$a_{ijt,t+2} = \frac{v_{ijt+1,t+2} - v_{ijt,t+1}}{s_{t,t+2}}$$

Of course acceleration must be computed on two contiguous time intervals $[t, t+1]$, $[t+1, t+2]$. In particular: $a_{ijt,t+2} \gtreqless 0$ $(a_{ijt,t+2} \lesseqgtr 0)$ if object $i$, for the $j^{(th)}$ variable, presents an increasing (decreasing) variation of velocity in the time interval from $t$ to $t+2$.

$a_{ijt,t+2} = 0$ if object $i$, for the $j^{(th)}$ variable, does not change velocity from $t$ to $t+2$.

Therefore, the basic information of a trajectory can by organized into three three-way matrices:

$\mathbf{X} = [\mathbf{X}_{..t} = [x_{ijt}]_{I \times J}, t = 1, ..., T]$
$\mathbf{V} = [\mathbf{V}_{..t,t+1} = [v_{ijt,t+1}]_{I \times J}, t = 1, ..., T-1]$
$\mathbf{A} = [\mathbf{A}_{..t,t+2} = [a_{ijt,t+2}]_{I \times J}, t = 1, ..., T-2]$

respectively for trend, velocity, and acceleration, where: $\mathbf{V}_{..t,t+1} = \frac{1}{s_{t,t+1}}(\mathbf{X}_{..t+1} - \mathbf{X}_{..t})$;
$\mathbf{A}_{..t,t+2} = \frac{1}{s_{t,t+2}}\left(\frac{1}{s_{t+1,t+2}}(\mathbf{X}_{..t+2} - \mathbf{X}_{..t+1}) - \frac{1}{s_{t,t+1}}(\mathbf{X}_{..t+1} - \mathbf{X}_{..t})\right)$

## 13.4    Dissimilarities between Trajectories

A dissimilarity between trends of objects $\mathbf{X}_{i..}$ and $\mathbf{X}_{l..}$ is evaluated according to a measure of distance between $\mathbf{X}_{i.t}$ and $\mathbf{X}_{l.t}$, for $t=1,...,T$:

$\delta(i,l) = \pi_1 \sum_{t=1}^{T-2} tr\left[(\mathbf{X}_{i.t} - \mathbf{X}_{l.t})'(\mathbf{X}_{i.t} - \mathbf{X}_{l.t})\right]$ where $\pi_1$ is a suitable weight to normalize distances.

A dissimilarity between velocities of objects $\mathbf{X}_{i..}$ and $\mathbf{X}_{l..}$ is evaluated according a measure of distance between $\mathbf{V}_{i.t,t+1} = (v_{i1t,t+1}, \ldots, v_{iJt,t+1})'$ and $\mathbf{V}_{l.t,t+1}, t = 1, \ldots, T-1$, in a time interval:

$\delta(i,l) = \pi_2 \sum_{t=1}^{T-2} tr\left[(\mathbf{V}_{i.t,t+1} - \mathbf{V}_{l.t,t+1})'(\mathbf{V}_{i.t,t+1} - \mathbf{V}_{l.t,t+1})\right]$ where $\pi_2$ is a suitable weight to normalize the velocity dissimilarity.

A dissimilarity between accelerations of objects $\mathbf{X}_{i..}$ and $\mathbf{X}_{l..}$ is evaluated according to a measure of distance between $\mathbf{A}_{i.t,t+2} = (a_{i1t,t+2}, \dots, a_{iJt,t+2})'$ and $\mathbf{A}_{l.t,t+2}$, $t = 1, \dots, T-2$:

$\delta(i,l) = \pi_3 \sum_{t=1}^{T-2} tr\left[(\mathbf{A}_{i.t,t+2} - \mathbf{A}_{l.t,t+2})'(\mathbf{A}_{i.t,t+2} - \mathbf{A}_{l.t,t+2})\right]$ where $\pi_3$ is a suitable weight to normalize the acceleration dissimilarity.

A dissimilarity between two trajectories that takes into account trend, velocity, and acceleration is thus formalized as the sum of the three individual dissimilarities:

$$d(i,l) = \pi_1 \sum_{t=1}^{T} ||\mathbf{X}_{i.t} - \mathbf{X}_{l.t}||^2 + \pi_2 \sum_{t=1}^{T-1} ||\mathbf{V}_{i.t,t+1} - \mathbf{V}_{l.t,t+1}||^2$$

$$+ \pi_3 \sum_{t=1}^{T-2} ||\mathbf{A}_{i.t,t+2} - \mathbf{A}_{l.t,t+2}||^2 \qquad (13.1)$$

## 13.5    The Clustering Problem

For clustering the trajectories we minimize the following loss function with respect to binary variable matrix $\mathbf{U}$, and continuous variables matrices $\overline{\mathbf{X}}_{..t}$, $\overline{\mathbf{V}}_{..t,t+1}$, and $\overline{\mathbf{A}}_{..t,t+2}$, where we add a feature of a dimensionality reduction of the variables via the orthonormal projection matrix $\mathbf{BB}'$.

$$\text{Min } \pi_1 \sum_{t=1}^{T} ||\mathbf{X}_{..t} - \mathbf{U}\overline{\mathbf{X}}_{..t}\mathbf{BB}'||^2 + \pi_2 \sum_{t=1}^{T-1} ||\mathbf{V}_{..t,t+1} - \mathbf{U}\overline{\mathbf{V}}_{..t,t+1}\mathbf{BB}'||^2$$

$$+ \pi_3 \sum_{t=1}^{T-2} ||\mathbf{A}_{..t,t+2} - \mathbf{U}\overline{\mathbf{A}}_{..t,t+2}\mathbf{BB}'||^2 \qquad (13.2)$$

subject to [P1]:

$\mathbf{B}'\mathbf{B} = \mathbf{I}_J$
$u_{ig} \in \{0,1\} \quad (i = 1, \dots, n; g = 1, \dots, G)$
$\sum_{g=1}^{G} u_{ig} = 1 \quad (i=1,\dots,n)$

where matrices $\overline{\mathbf{X}}_{..t}$, $\overline{\mathbf{V}}_{..t,t+1}$, and $\overline{\mathbf{A}}_{..t,t+2}$ are the matrices of the G consensus trajectories including trend, velocity, and acceleration information.

In problem [P1] the observed trajectories are classified in G consensus trajectories and their location in the space is identified. Furthermore, we suppose we consider a dimensionality reduction specified by the orthonormal projection matrix $\mathbf{BB}'$.

The quadratic problem [P1] in the continuous variables $\overline{\mathbf{X}}_{..t}$, $\overline{\mathbf{V}}_{..t,t+1}$, and $\overline{\mathbf{A}}_{..t,t+2}$ and binary $\mathbf{U}$ is solved here by using the sequential quadratic algorithm (SQP) [260]. It is well known that the partitioning of $n$ objects in $k$ clusters is a non-polynomial (NP)-hard problem in the class of the NP-complete problems [190], therefore the problem of clustering trajectories that is a three-way extension is also NP-hard and no guarantee to find the optimal solution is available. Therefore, to increase the chance to find the optimal solution a multistart procedure is applied that consists of starting the algorithm from different random solutions and retaining the best solution.

## 13.6   Application

Cancer mortality data, initially from 122 countries, were extracted from the World Health Organization statistical database (WHOSIS) in March 2005. This database contains absolute numbers of deaths officially reported by WHO member states for the years 1980–2000. For these years, the WHO database includes cause-of-death statistics, coded according to two former versions of the ICD (International Classification of Diseases, version 9 from 1979 to 1998, and version 10 from 1999); the years of transition between the two versions exhibit large differences among countries. We accounted for these changes in disease classification by using specific transition datasets available on the WHO Web site since 2005. In this chapter we focus on lung cancer, because the mortality data were available across a sufficient number of countries, age bins, and years. Data related to the age below five (especially for the age below one year) and above 89 are reported in a heterogeneous and incomplete way across countries. In the present analysis, only age groups from 40 to 74 were considered (leading to seven 5-year age bins: 40–44, 45–49, 50–54, 55–59, 60–64, 65–69, 70–74). To account for differences in country- and period-specific variations in age distributions, the mortality data with respect to lung cancer were directly age-standardized according to Segi's world population [137], for men.

We exclude 23 countries because there were clearly visible outliers from the local (i.e., country-specific) trend (e.g., Brazil 1971, 1982; Chile 1988; Portugal 1988, 1989). Moreover, in order to assess the degree of annual variation of the data, coefficients of variation were computed, and the countries above the 80th percentile were excluded.

Information on mortality for the period of 1980 to 2000 was missing for more than five years in 47 countries. Therefore, these countries were not considered in the present analysis. For the remaining 51 countries with less than five years of missing data, imputation was undertaken by interpolation (spline interpolation when possible, otherwise linear interpolation).

For the years up to 1998, the lung cancer absolute frequencies were provided in the WHO database, whereas for the years from 1999 (ICD 10) they had to be computed by integrating across absolute frequencies for all specific sites. We aim at categorizing the evolution of lung cancer mortality in the past 21 years from 1980 to 2000 in the selected 51 countries; and we expect to uncover some general trends in the clusters.

The resulting data array is a three-way (51,21,7) table, 51 countries, 21 years, 7 age bins.

We present some results for partitions from 2 up to 11 clusters, and some more detailed results for the 2-cluster partition and for the 11-cluster partition. The algorithm indicates the 2-cluster partition as the optimum one according to the pseudo F criterion ("Optimal F" concerns the final reduced space) as Table 13.1 shows.

**TABLE 13.1**

Partitions and their properties.

| Clusters | Pseudo F | Optimal F |
|---|---|---|
| 2 | 238.6807 | 27.2302 |
| 3 | 183.8518 | 24.2847 |
| 4 | 146.6301 | 23.0354 |
| 5 | 124.0362 | 21.114 |
| 6 | 107.8434 | 21.114 |
| 7 | 96.334 | 20.3309 |
| 8 | 85.9706 | 19.9132 |
| 9 | 77.8542 | 19.5201 |
| 10 | 71.4224 | 19.1255 |
| 11 | 67.3236 | 18.4785 |

Figure 13.2 shows the consensus trajectories for the 2-cluster partition.

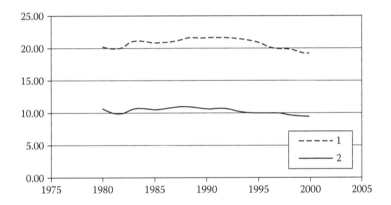

**FIGURE 13.2**

The consensus trajectories for the 2-cluster partition.

Figure 13.3 represents the two consensus trajectories in the first factorial plane.

Figure 13.4 shows the eleven consensus trajectories of the 11-cluster partition on the first principal component whereas Figure 13.5 selects the less "erratic" ones: three clusters, cluster 4, cluster 7, and cluster 11 appear to be sparse (less than two elements) and could be considered as outliers (countries Kuwait – cluster 4, Estonia – cluster 7, Trinidad and Costa Rica – cluster 11).

Figure 13.6 shows a geographical map restricted to Europe with the country memberships in the selected eight clusters.

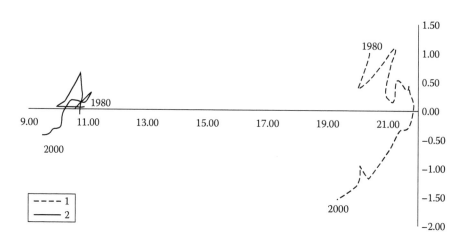

**FIGURE 13.3**
The two consensus trajectories in the first factorial plane.

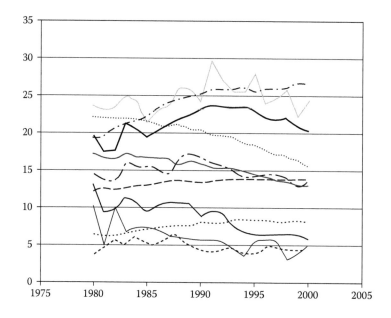

**FIGURE 13.4**
11-cluster partition.

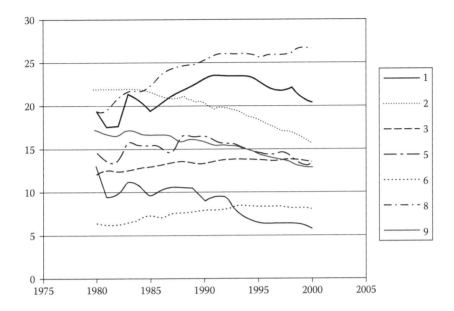

**FIGURE 13.5**
Eight consistent clusters out of the 11-cluster partition.

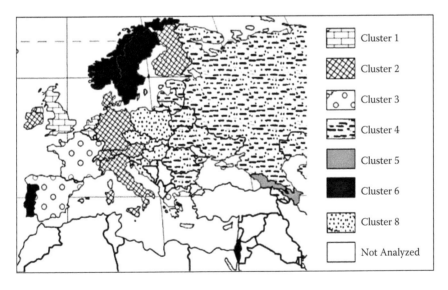

**FIGURE 13.6**
Eight clusters' projection in Europe.

## 13.7   Conclusions

We still have a lack of theoretical elements for solutions during the approximation step, particularly in the weighting step for the different components in the distance formula. We are evaluating a coefficient based on the Mahalanobis distance, but it needs to be adjusted at each step and a global satisfying procedure has not yet been found because sequential quadratic programming is quite computationally demanding. A more specific algorithm is needed and we are studying a new version of a fast coordinate descent algorithm. The results of this application on the lung cancer data can be compared to those of more complete studies on cancer evolution typologies [204][142]. Some convergence can be pointed out and particularly the similarity between the 8-cluster partitions (mainly for European countries), even if not exactly identical everywhere. Nevertheless, we also observe some discrepancies:

- Geographical proximities appear to be respected by the clustering procedure when it deals only with the values, but they are less apparent when including velocity and acceleration (e.g., Hong Kong is associated with Australia, but also with Austria).

- The 2-cluster partition was expected to group countries according to "western style of life" (with a convergent decreasing pattern on values) as opposed to the complementary group of countries; we do not obtain the same finding in the new approach (for example France is associated with Turkmenistan). It seems that the partition into two clusters is mainly linked to the different levels of the variables. Probably for this partition, values are more important than contiguity of countries.

Further work should focus on the elaboration of interpretation procedures for the resulting clusters in the approach presented in this chapter. How can the axes of Figure 13.3 be labeled in order to reveal why cluster 2 has a larger range of coordinates on the second axis than cluster 1 along the 21 years? What are the main effects on the groupings: common ranges both for values and velocities, or whatever else? Moreover, supplementary (illustrative) variables could be introduced in order to propose hints for explaining the between-cluster differences. This approach, which requires optimization algorithms in order to be efficient, is able to group trajectories according to their shape and their values: appropriate choices of weights in the dissimilarity formulas give more importance either to the values or to the shape in the clustering process.

## Acknowledgments

This work was supported by the ISTHMA Company that provided the cancer database.

# 14

## Trees with Soft Nodes: A New Approach to the Construction of Prediction Trees from Data

Antonio Ciampi

## CONTENTS

## 14.1   Introduction

A prediction tree is a structure of the type illustrated in Figure 14.1. The tree is shown "upside down," with the root at the top and the branches pointing downwards to the leaves. The graph shows how a prediction is made about a dependent variable (the name of a color, in the case of the figure), given the values of certain independent variables called predictors (the $z$'s). The prediction may take the form of a classification: if $z_1 < 4.8$, then the observation is green; or of a probability distribution: if $z_1 < 4.8$, then the observation is *green* with probability 0.60, *black* with probability 0.4, and *red* with *probability 0*. The figure also shows that each node of the tree induces a partition of the predictor space, hence the name of "recursive partition" often used to denote algorithms of tree construction from data.

There is a great variety of algorithms for constructing prediction trees from data, the most popular of which is CART, acronym for Classification And Regression Trees) [1]. All such algorithms share a central idea: at each step of a recursive procedure, a subsample of the data is partitioned according to the value of a predictor, the actual split being chosen so as to maximize a measure of split quality. There are numerous variants to this basic approach; we mention here only two basic choices that create important differentiation among algorithms: 1) the choice of the goodness-of-split measure; and 2) the choice of a procedure for determining tree size. In our previous work we have used the term generalized information to denote any measure of quality of a partition. Indeed, the term *information* is very appropriate in the particular case where data are adequately represented by a parametric probabilistic model: in this case it has been shown that the likelihood ratio statistic measures the information we can acquire about the dependent variable if we study it separately on each set of the partition [3]. The idea of information is also central to Quinlan's work

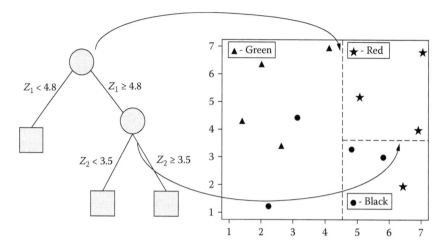

**FIGURE 14.1**
A prediction tree.

(Quinlan 1993 [264]), author of one of the most influential tree-growing algorithms in machine learning. The success of trees in data analysis is due to a great extent to the simplicity of the prediction rule they describe: indeed, the user arrives at a direct prediction of the dependent variable by answering a number, usually quite small, of binary questions, each concerning a single predictor. On the other hand, if the main goal of the user is to obtain an accurate prediction, even at the cost of reduced interpretability, then there is ample choice of algorithms of potentially higher predictive performance than prediction trees, such as feed-forward neural networks, support vector machines, and even modern forms of adaptive regression. Moreover, trees and other algorithms can be improved upon by so-called ensemble approaches. These approaches use re-sampling techniques to obtain a family of predictors appropriately combined to produce a super predictor, often of exceptional accuracy. Perhaps the most popular ensemble prediction technique is known as random forest, since it consists of a large number of trees acting cooperatively to make predictions, given predictors (Breiman 2001 [56]). Therefore, one might be tempted to conclude that prediction trees are outdated and that one must accept some loss of interpretability, in exchange for the superior performance of the new methodologies. Our point of view differs, as we are convinced that tree construction can still be enhanced while keeping a good dose of interpretability. Specifically, we propose to radically transform the recursive partition algorithm, which is inherently local, by making it global, i.e., by using all data, at each step of the construction. Nodes still play an essential role, but, as we shall see, branching is replaced by weighting. The next section traces the main idea of this work to certain development of symbolic data analysis (Billard & Diday 2006 [24]). In Sections 14.3 and 14.4 we introduce the notion of soft node, and tree with soft nodes respectively. Two examples of real data analysis are presented in Section 14.5. Section 14.6 summarizes the results of some comparative empirical evaluation of prediction trees with soft nodes. Section 14.7, is devoted to a discussion and to current and future research directions.

## 14.2 Trees for Symbolic Data

A standard tree is constructed from a "numerical" data matrix, i.e., from a matrix consisting of numerical entries that represent values taken by variables on observations. A standard tree, such as the one in Figure 14.1, may serve as a rule for attaching a value of the dependent variable to any observation, in particular to a new one, given only the values of its predictors. What if we wish to make a prediction for an observation given its "symbolic" predictors (Billard & Diday 2006 [24])? In this chapter we will only consider a particular type of symbolic data: imprecise data with imprecision represented by a range of possible values. For example, a specific observation could be: $z* = (z_1, z_2) = ([4.5, 5.0], [3.4, 3.9])$, which is equivalent to saying that the value of $z_1$ is between 4.5 and 5.0, and that the value of $z_2$ is between 3.4 and 3.9. Suppose that $z*$ is observed on a unit drawn from the same population used to construct the tree of Figure 14.1. How can we classify such an observation? Quinlan (Quinlan 1990 [263]) suggested that, at each node, the unit should be sent partially to the right and partially to the left. Specifically, at a node defined by a constraint on the variable $z_i$, one should send to the left a fraction of the observation and to the right the complementary fraction; the fraction sent to the left should be calculated as the ratio of the length of the subinterval satisfying the constraint to the total length of the interval. We can, of course, interpret this as a probabilistic decision under the implicit assumption that the (real but unknown) value of $z_i$ is uniformly distributed on the interval in which it ranges. This is illustrated by Figure 14.2.

Clearly, it is not difficult to use a standard tree, constructed from numerical data, to classify a symbolic observation. We may now ask: is it possible to construct a standard tree from symbolic data? We have answered this question in the affirmative in (Ciampi et al. 2000 [72]). Without entering into a detailed explanation of the proposed solution, let us look at the artificial example contained. Table 14.1 shows a small data matrix representing symbolic data: we have here 12 observations, each belonging without ambiguity to one of three possible classes, but with values of the predictors known only imprecisely, in that they are only known to range in a given interval.

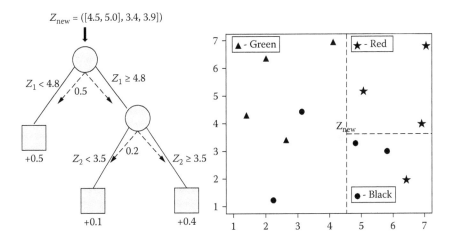

**FIGURE 14.2**
Classifying an imprecise data point (Quinlan, 1990).

**TABLE 14.1**
Example of symbolic data.

| $z_1$ | $z_2$ | Class |
|---|---|---|
| $[1.0, 3.0]$ | $[1.5, 2.0]$ | 1 |
| $[2.5, 3.5]$ | $[3.0, 5.0]$ | 1 |
| $[3.5, 6.5]$ | $[3.0, 3.5]$ | 1 |
| $[5.0, 7.0]$ | $[1.5, 4.5]$ | 1 |
| $[4.0, 4.8]$ | $[0.5, 2.0]$ | 2 |
| $[7.0, 7.5]$ | $[2.5, 5.0]$ | 2 |
| $[7.0, 8.0]$ | $[5.5, 6.5]$ | 2 |
| $[4.0, 6.5]$ | $[4.0, 5.5]$ | 2 |
| $[3.0, 6.0]$ | $[6.0, 6.5]$ | 3 |
| $[0.5, 1.5]$ | $[3.0, 5.0]$ | 3 |
| $[1.5, 2.5]$ | $[5.5, 6.0]$ | 3 |

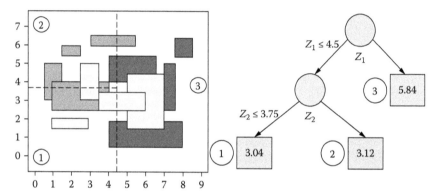

**FIGURE 14.3**
A "hard" tree constructed from imprecise data.

These data are represented graphically in Figure 14.3, left; Figure 14.3, right, represents the tree constructed from these data using the algorithm described in (Ciampi et al. 2000 [72]).

Finally we may ask a further question: is it possible and useful to construct from numeric data a tree with probabilistic decisions at its nodes? In what follows we will provide an affirmative answer to this question also.

## 14.3  Soft Nodes

In what follows we will use the terms *hard node* and *hard partition* to refer to a standard node and a standard partition, respectively. We will use the qualifier soft in the context of probabilistic nodes and partitions. A soft node represents a binary probabilistic decision rule: "At the $i$-th node, take the LEFT path with probability $p(z_i)$ AND the RIGHT path with the complementary probability $q(z_i) = 1 - p(z_i)$." It is therefore the soft node function $p(z_i)$ that summarizes the decision rule. To construct this function from the data, we have to simply let the data select a function from a flexible and appropriate class. In this chapter we limit ourselves, for convenience, to a class of nondecreasing sigmoid functions with values

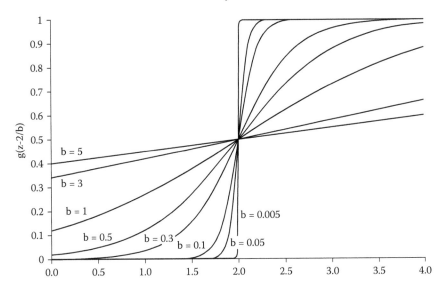

**FIGURE 14.4**
The logistic function for several values of the parameters "cut" and "softness."

comprised between 0 and 1. In principle, a soft partition may well be preferable to a hard one, i.e., it may result in a larger information gain about the dependent variable than the best hard partition. In fact, it is easy to see that the best soft partition cannot be worse than the best hard partition in terms of information gain, for a hard node can be represented by the indicator function $I[z, a]$, of the interval $[a, +\infty)$, and this can be expressed as the limit of nondecreasing sigmoid functions. To choose the cut point associated with a hard node one needs to determine from the data only the cut point $a$. In contrast, the choice of the soft node function may require several parameters, depending on the flexibility of the chosen family. To further simplify the problem, in this chapter we limit ourselves to model the function $p(z)$ by the 2-parameter logistic family, defined as:

$$p(z; a, b) = g(\frac{z - a}{b})$$

with:

$$g(z) = \frac{1}{1 + e^{-z}}$$

Figure 14.4 illustrates the flexibility of the logistic family. We will refer to the parameters $a$ and $b$ as cut point and softness, respectively. Clearly, a hard node has softness equal to zero.

When predicting a binary variable $y$ (two classes), we can calculate the empirical information gain associated with the node $N$ by comparing two prediction models: the trivial model, $p = P[y = 1] = 1$, and the model associated with the soft node function:

$$p(z) = p_{left}g(\frac{z - a}{b}) + p_{right}(1 - g(\frac{z - a}{b})) \tag{14.1}$$

Then the information gain associated with the node $N$ is estimated by the Likelihood Ratio Statistic (LRS) of the comparison (empirical information gain). It is interesting to note that the second model represents the distribution of $y$ as a mixture of Bernoulli distributions.

## 14.4    Trees with Soft Nodes

At a soft node all data are used to estimate the parameters of the soft node function (Breiman et al. 1984 [57]). In contrast, the estimation of the parameters $p_{left}$ and $p_{right}$ associated with a hard node is strictly local. It follows that in a hard tree the prediction parameters are estimated on samples of sizes that are rapidly decreasing as the tree grows in depth: this is one of the leading causes of high prediction variability and structural instability of standard trees. On the other hand, a soft tree is grown using an algorithm that uses all the data to estimate all the parameters. This increases the capability of borrowing strength from other data, a very prized feature in statistical modeling, which underlies, for example, the great success of ubiquitous regression models. We recall that the prediction model associated with a hard tree has the form:

$$p(z) = p_1 I_1(z) + p_2 I_2(z) + \dots p_L I_L(z) \tag{14.2}$$

where the $I_k, k = 1, 2, \dots L$, denote the indicator function of the $k$th leaf of the tree, and $p_k$ the probability that $y = 1$ for an observation belonging to the $k$th leaf. The $I$s are, of course, products of the indicator functions corresponding to the questions associated with the nodes of the tree. The key idea of this work is to define a soft prediction tree by substituting the $I$s of the equation with products of soft node functions denoted by $J$:

$$p(z) = p_1 J_1(z) + p_2 J_2(z) + \dots p_L J_L(z) \tag{14.3}$$

Once the soft tree $T$ is defined, it is easy to associate with it an *information gain $IC(T)$* as the likelihood ratio for comparing the model associated with $T$ to the trivial model ($p = P[y = 1] = constant$). In practice, this gain is estimated by the corresponding LRS. Rather than developing a general formalism, here we will limit ourselves to an example that illustrates equation 14.3. In Figure 14.5, we represent a tree with two nodes, one hard and one soft. SBP is the acronym of Systolic Blood Pressure. Note that a node defined by a binary predictor cannot be soft, because of an identification problem (see (Ciampi et al. 2002 [71])).

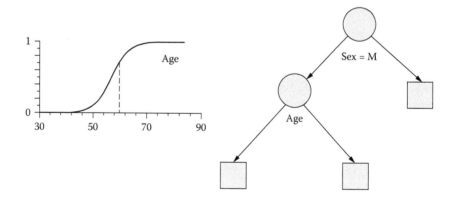

**FIGURE 14.5**
A soft tree.

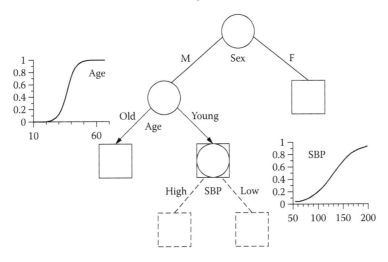

**FIGURE 14.6**
Construction of a soft tree from data.

Now, the prediction model associated with this tree is:

$$
\begin{aligned}
p(Sex, Age) &= p_1 I[Sex = M]g(Age \geq 65) \\
&+ P_2 I[Sex = M](1 - g(Age \geq 65)) \\
&+ P - 3I[Sex = F]
\end{aligned}
\tag{14.4}
$$

A simple algorithm of tree construction can be developed from the following statement: *Enlarge the current tree by adding a branch in such a way as to maximally increase the empirical information gain.*

The algorithm used in (Ciampi et al. 2002 [71]), was in fact based on the above, with a stopping rule determined by an appropriate criterion, such as the Akaike information criterion (AIC) or the Bayesian information criterion (BIC) (Burnham & Anderson 2002 [62]). To compute the information gain and the stopping criterion at each step of the construction, we need to maximize the log-likelihood of the current tree and of the enlarged tree. Since the basic model is a mixture of distributions, our maximization algorithm is based on the expectation maximization (EM) type. Therefore, we do not directly maximize the mixture model, but we aim to maximize the log-likelihood of the complete data, i.e., based on hypothetical data similar to ours but augmented by the unobserved class (distribution) indicator for each observation. For the computation to be possible, we have to replace the unobserved indicators by estimates based on the data we actually observe. In fact, the algorithm alternates until convergent between an E-step (E for expectation) and an M-step (M for maximization). In the E-step we replace the class indicators by their conditional expectation given the observed data; in the M-step we maximize the complete data log-likelihood, as updated by the previous E-step. The algorithm is described in (Ciampi et al. 2002 [71]). As an example, we write here the log-likelihood of the prediction model given by the tree of Figure 14.6 (before adding the dotted branch). Since this tree has only one soft node, the equation contains only one class indicator denoted by $\zeta_i$. We have:

$$
\begin{aligned}
l(\theta) \;=\; & \sum_{i=1}^{n} \{ y_i \zeta_i I[sex_i = M] log(p_1) + (1 - y_i)\zeta_i I[sex_i = M] log(1 - p_1) \\
& +\; y_i(1 - \zeta_i)I[sex_i = M]log(p_2) + (1 - y_i)(1 - \zeta_i)I[sex_i = M]log(1 - p_2) \\
& +\; y_i I^c[sex_i = M]log(p_3) + (1 - y_i)I^c[sex_i = M]log(1 - p_3) \} \\
& +\; \sum_{i=1}^{n} \{ \zeta_i log(g((age_i - a)/b)) + (1 - \zeta_i)log(g^c((age_i - a)/b)) \}
\end{aligned}
$$

## 14.5   Examples

The first example concerns the development of a tree-structured prediction rule for Type II diabetes from well-known risk factors. We show here a soft and a hard tree constructed from a public domain database, containing data on 523 women of the American Indian culture Pima, living in the region of Phoenix, Arizona. We used the following predictors (their abbreviation in parentheses):

- Glucose (Glu)

- Diastolic Blood Pressure (bp)

- Triceps skinfold thickness (skin)

- Body Mass Index (BMI)

- Diabetes Pedigree Function (Ped)

- Age.

The hard and soft tree constructed from these data are shown in Figure 14.7, top and bottom panels, respectively.

The soft tree has a slightly smaller prediction error (19.1%) than the hard tree (21%). Also, the prediction rule associated with the soft tree is simpler than that of the hard tree. On the one hand, the hard tree has nodes including a fewer number of variables (Glucose, Age, BMI), but each variable appears twice, suggesting that hard nodes are not very appropriate. On the other hand, the soft tree contains more variables (Glucose, Age, and Pedigree Function) but fewer repetitions (BMI appears twice). As a second example, we show a classification problem with 4 classes, i.e., the dependent variable is 4-level categorical. Relatively simple modifications were needed to adapt the algorithm of tree construction to the case of $k$-level categorical dependent variable ($k \geq 2$). The most important modification is to rewrite equation 14.3 with vector parameters instead of scalar parameters.

We show here the tree analysis of the public domain dataset CRABS. The goal is to classify a particular crab into one of four classes: Blue Male (BM), Blue Female (BF), Orange Male (OM), and Orange Female (OF). The predictors are the following morphological characteristics: Frontal Lobe size (FL), Rear Width (RW), Carapace Length (CL), Carapace Width (CW), and Body Depth (BD). The soft tree obtained from our algorithm is given in Figure 14.8.

On the whole, the view on the data offered by this soft tree is quite simple and similar to what we would see if the nodes where hard: a) Rear Width separates the Orange Female from

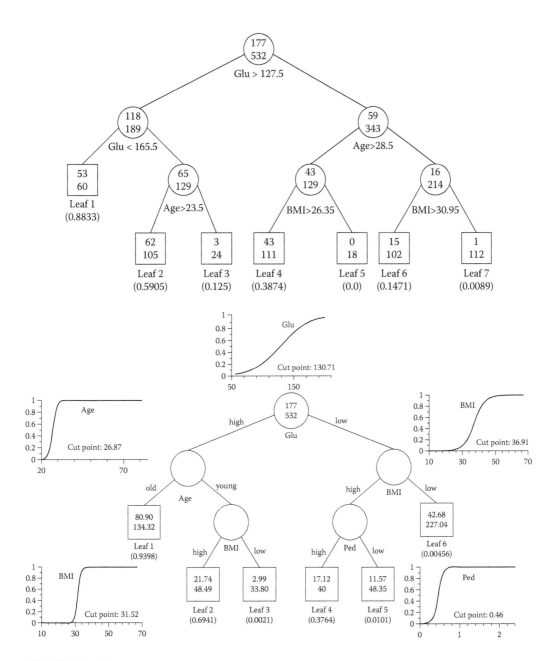

**FIGURE 14.7**
Top: PIMA data, hard tree, bottom: PIMA Data, soft tree.

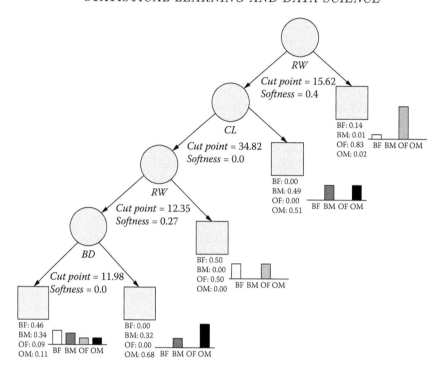

**FIGURE 14.8**
CRABS Data: soft tree.

the rest; b) among the crabs with small RW, Males have a longer carapace than Females; c) Blue Crabs tend to be smaller than Orange Crabs; and d) in general, it is harder to discriminate between sexes than between colors. On the other hand, the classification rule associated with the soft tree is slightly more complex than the one associated with the corresponding hard tree. For instance, consider crab #134 (an Orange Male); its measures are: $FL = 18.4, RW = 13.4, CL = 37.9, CW = 42.2, BD = 17.7$. At the root node, crab #134 is sent to the left with probability 0.996; the second node being hard (zero softness), it goes right with probability 1. As a result, this crab is sent with high probability to the second leaf from the top, where it is correctly classified by the majority rule as being OM.

## 14.6  Evaluation

We present now an evaluation of the soft tree based on the approach by Breiman (Breiman 1996 [55]). We summarize here the results for a binary dependent variable, see (Ciampi et al. 2002 [71]) for further details. We have selected 6 public domain databases described in Table 14.2. For every database we have repeated the following steps 100 times:

1. Randomly select 10% of the data, for use as Test Set (TS), while the rest of the data is used as Learning Set (LS).

2. From the LS, build a soft tree and a hard tree.

3. On the TS, evaluate the performance of each tree by: the misclassification error, the

**TABLE 14.2**
Characteristics of the 6 databases used for the evaluation *C = continuous variable, D = discrete variable.

| | Size of Original Sample | Size of Useable Sample | Sample Size of Test Set | Number of Variables |
|---|---|---|---|---|
| Breast Cancer | 683 | 683 | 68 | 9 (C = 9; D = 0)* |
| PIMA | 532 | 532 | 50 | 7 (C = 7; D = 0) |
| Heart disease | 303 | 296 | 30 | 8 (C = 5; D = 3) |
| Liver disease | 345 | 345 | 35 | 5 (C = 5; D = 0) |
| Type II Diabetes | 403 | 367 | 37 | 8 (C = 6; D = 2) |
| Prostate cancer | 502 | 482 | 48 | 13 (C = 10; D = 3) |

c-index [equal to the area under the receiver operating characteristic (ROC) curve], the Brier score, the deviance, and the number of leaves.

The results are summarized in Table 14.2.
We remark the following:

a) For all datasets, with the exception of Type II Diabetes, the soft tree has a smaller misclassification error than the hard tree, the improvement varying between 4% and 13%.

b) According to the other criteria (data on the Brier score not shown), the soft tree performs uniformly better than the hard tree. Indeed, the soft tree has a smaller deviance, smaller Brier, score, and larger c-index than the hard tree, for all the datasets in Table 14.2.

## 14.7 Discussion

We have presented here the basic concepts underlying trees with soft nodes, and have outlined an algorithm to construct such trees from data. The key concept is the probabilistic (soft) node. On the one hand, a probabilistic node can be seen as an artificial construct useful to obtain a better decision than the one provided by a hard decision node. On the other hand, in some specific cases, a probabilistic node may well be an adequate representation of a latent but very real binary variable, for example, the indicator of pathogenic variant of a gene. The empirical results presented in this chapter indicate that a tree with soft nodes may have a higher predictive performance than a hard tree, while still providing an appealing interpretation. Unlike a hard tree, a soft tree defines homogeneous groups in a way that is more qualitative than quantitative; for example, a leaf of a soft tree may be defined by "very high values" of certain variables, and by "very low values" of certain others. In general, a leaf represents an "extreme case" or stereotype, and every observation is described as a mixture of these stereotypes. At the same time, a soft tree can be seen as a rule for making individual predictions. As our results show, this rule seems to be of good quality. This double feature of a global interpretation that is less crisp and an individual prediction that is more complex may appear at first sight less desirable than the simplicity of a hard tree. The latter offers a model that also works at the global and at the individual level: it specifies quite precisely the expression "high value" in terms of a cut point, and assigns an observation to a unique leaf.

We note also the soft trees tend to have more leaves and less repetitions of one predictor at different nodes. This is likely to be a consequence of the greater flexibility of a soft node as compared with a hard node, and of the fact that each observation contributes, though with varying weights, to each step of the soft tree construction. For the soft tree to be adopted in the common practice of data analysis, one needs to demonstrate that the loss in simplicity is offset by the gain in realism and predictive accuracy. It is also necessary to investigate the scalability of the algorithm, with a view towards application to much larger datasets than those treated so far. The extent of the required effort is not yet known. In our current research we are addressing these problems. In particular, we are extending simulations and comparative evaluation of the soft tree to the case of categorical ($k > 2$) and quantitative dependent variables. Also, we are studying ways to improve the EM algorithm exploiting specific features of our problem. As for future developments, we plan to extend the soft tree approach to the prediction of complex dependent variables, such as censored survival times and multivariate continuous and discrete responses. In conclusion, we remark that our research itinerary has lead us successively from the construction of hard trees from numeric data, to the construction of hard trees from symbolic data, and from this to the notion and construction of soft trees from numeric data. It would be interesting and probably useful to advance towards the development of soft trees from symbolic data.

# 15

## Synthesis of Objects

**Myriam Touati, Mohamed Djedour, and Edwin Diday**

## CONTENTS

## 15.1  Introduction

The aim of the symbolic approach in data analysis is to extend problems, methods, and algorithms used on classical data to more complex data called "symbolic", which are well adapted to represent clusters, categories, or concepts (Diday 2005 [97]). Concepts are defined by intent and an extent that satisfy the intent. For example, the swallow's species can be considered as a concept, the intent is the description of this species of birds, and the set of swallows is the extent of this concept. Symbolic data are used in order to model concepts by the so called "symbolic objects". We mention the definition of three kinds of symbolic objects: events, assertions, and synthesis objects. We focus on synthesis objects. For example having defined vases by their color, size, material, etc., and flowers described by their species, type, color, stem, etc., our aim is to find "synthesis objects", which associate together the most harmonious vases and flowers. This approach can also be applied in the fashion domain in

171

order to improve clothes of a mannequin, or in advertizing in order to improve the objects represented in a poster. More formally, we consider several objects characterized by some properties and background knowledge expressed by taxonomies, similarity measures, and affinities between some values. With all such data, the idea is to discover a set of classes, which are described by modal synthesis objects, built up by assertions of high affinity. We use a symbolic clustering algorithm that provides a solution allowing for convergence towards a local optimum for a given criterion and a generalization algorithm to build modal synthesis objects. We give finally an example to illustrate this method: the advising of the University of Algiers students.

Symbolic Data Analysis (SDA) was born from several influences and mainly by data analysis (Benzécri et al. 1973 [17]) and artificial intelligence (Michalski et al. 1981 [230]), and its first basic ideas can be found in (Diday 1988 [95]). Symbolic Data Analysis is an extension of classical data analysis and seeks to treat more complex elements that contain variability such as images, texts, and more generally hierarchical data, related data tables (such as the ones contained in relational data bases), time series, etc.

The usual data mining model is based on two parts: the first concerns the observations also called individuals in this paper, and the second contains their description by several standard variables including numerical or categorical. The Symbolic Data Analysis (Bock & Diday 2000 [29]), (Billard & Diday 2006 [24]), (Diday & Noirhomme-Fraiture 2008 [100]) model needs two more parts: the first concerns clusters of individuals modeled by concepts, and the second concerns their description. The concepts are characterized by a set of properties called intent and by an extent defined by the set of individuals that satisfy these properties. In order to take into account the variation of their extent, these concepts are described by symbolic data, which may be standard categorical or numerical data, but moreover are multivalued, intervals, histograms, sequences of weighted values and the like. These new kinds of data are called symbolic as they cannot be manipulated as numbers. Then, based on this model, new knowledge can be extracted by new tools of data mining extended to concepts considered as new kinds of individuals. Symbolic analysis has been successfully used in government statistics, species data, social sciences, and the like. There are many examples from a wide range of fields in (Billard & Diday 2006 [24]).

In this chapter, the aim is to extend and adapt a method and an algorithm of classical data analysis to symbolic objects such as "assertions" which are not necessarily defined on the same variables, using the "affinities" to classify clusters and the notion of modal symbolic objects to express the results. In this chapter, first, we define the nature of symbolic objects, "events" and "assertions" that constitute the input data (of the algorithm explained in the second part of the paper), the output data ("synthesis objects" and "modal synthesis objects") and the additional knowledge, as in the machine learning domain, in the form of "affinities", measures of similarity and taxonomies. We then propose an algorithm of the "dynamic clustering" type adapted to symbolic objects in order to obtain clusters of "assertions" grouped by high "affinity". We describe in the next step an algorithm of generalization of the obtained clusters in order to build "synthesis objects".

Finally, we illustrate this method by an application used to advise students from the University of Algiers, where assertions of input are academic sectors and the students described by their Baccalaureate results and where "affinities" are calculated according to a scale determined by an expert. The goal is to define simultanously students and academic sector with which the students have good affinities.

## 15.2 Some Symbolic Object Definitions

Symbolic objects are defined by a logical conjunction of elementary "events" where values or variables may be single, multiple, or nonexistent and where objects are not necessarily defined on the same variables. Let us recall a few symbolic object definitions that have been given in (Bock & Diday 2000 [29]).

Let's consider a finite set $\Omega$ of elementary objects and the set of variables $\{y_i : \Omega \to \mathbf{O}_i\}_{i=1,...,n}$. In Figure 15.1 drawing by E. Diday, the set of individuals and the set of concepts constitute the real-world sets. We might say that the world of objects in the real world – the individuals and concepts – are considered as lower- and higher-level objects. The modeled world is composed of the set of descriptions that models individuals (or classes of individuals) and the set of symbolic objects that models concepts. The starting point is a given class of individuals associated with a concept $C$. If the concept $C$ is the swallow species, and 30 individual swallows have been captured among a sample $\Omega'$ of the swallows of an island, $Ext(C/\Omega')$ is known to be composed of these 30 swallows. Each individual $\omega$ in the extent of $C$ in $\Omega'$ is described in the description space by using the mapping $y$ such that $y(\omega) = d_\omega$ describes the individual $\omega$. Given a class of 30 similar birds obtained, for example, from a clustering process, a concept called swallow is associated with this class (the dotted ellipse containing the 3 x's). The concept $C$ is modeled in the set of symbolic objects by the following steps:

First, the set of descriptions of the individuals of $Ext(C/\Omega')$, or the class of individuals (in the dotted ellipse) associated with the concept $C$, is generalized with an operator $T$ (as it can be a T-norm) in order to produce the description $d_C$ (which may be a set of Cartesian products of intervals and/or distributions or just a unique joint distribution, etc.).

Second, the matching relation $R$ can be chosen in relation to the choice of $T$. The membership function is then defined by the mapping $a_C : \Omega \to L$ where $L$ is $[0,1]$ or $\{true, false\}$ such that $a_C(\omega) = [y(\omega)R_C d_C]$, which measures the fit or match between the description $y(\omega) = d(\omega)$ of the unit $\omega$ and the description $d_C$ of the extent of the concept $C$ in the database. The symbolic object modeling the concept $C$ by the triple $s = (a_C, R, d_C)$ can then be defined. Knowing the extent $E_1$ of a concept in $\Omega'$ and the extent $E_2$ in $\Omega'$ of a symbolic object that models this concept, $T, R$ and $a$ have to be chosen such that $E_1$ and

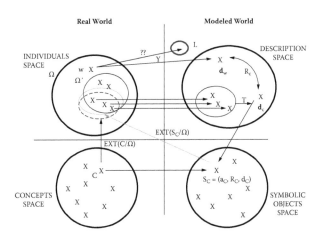

**FIGURE 15.1**
Modeling by a symbolic object of a concept known by its extent.

$E_2$ become as close as possible. Numerous learning processes and their convergence can be defined depending on the kind of organization or modeling of the concepts. Diday & Emilion (1997, 2003) [98, 99] showed that as the number of individuals increases, the concepts and their Galois lattice structure converge.

**Definition 1** *A **symbolic object** s is the triple $s = (a, R, d)$ where a is a mapping: $\Omega \to L$, which maps individuals $\omega \in \Omega$ onto the space L depending on the relation R between its description and the description d. When $L = \{0, 1\}$, i.e., when a is a binary mapping, then s is a **Boolean** symbolic object. When $L = [0, 1]$, then s is a **modal** symbolic object.*

Expressed in the form of a combination of properties, the symbolic object $s$ was defined in its intent.

The extent of a symbolic object $s$, denoted by $Ext(s)$, is the set of all individuals $\omega$ from $\Omega$ with $a(\omega) = true$. The extent of $s$ in $\Omega$ is defined as follows:

$$Ext(s) = \{\omega \in \Omega / y_i(\omega) \in V_i\} = \{\omega \in \Omega / a(\omega) = true\}.$$

Let $s_1$ and $s_2$ be two symbolic objects. We say that $s_1$ inherits from $s_2$ and that $s_2$ is more general than $s_1$ if and only if $Ext(s_1) \subset Ext(s_2)$.

**Example 1** *Suppose we have a symbolic object s where the intent is:*
$s = [size \in [20, 40]] \wedge [color \in \{blue, white\}]$.
*A vase belongs to the extent of s if its size is between 20 and 40 cm and if its color is blue or white.*

**Definition 2** *An **event** $ev_i$ is a symbolic object defined by the triple $ev_i = (e_i, R_i, V_i)$ where $e_i$ is the binary mapping: $\Omega \to \{true, false\}$ defined by the logical expression:*

$$e_i = [y_i R_i V_i]$$

*such that $e_i(\omega) = true$ if and only if the logical proposal "$y_i(\omega) R_i V_i$" is true.*

**Example 2** *$ev_1 = (e_1, \in, [20, 30])$ where $e_1 = [size \in [20, 30]]$ and $e_1(\omega) = true$ if the size of $\omega$ is between 20 and 30;*
*$ev_2 = (e_2, \in, \{black, white, red\})$ where $e_2 = [color \in \{black, white, red\}]$ and $e_2(\omega) = true$ if the color of $\omega$ is either black or white or red.*

**Definition 3** *An **assertion** a is a symbolic object defined by the triple $a = (a_j, R_j, V_j)$ where $a_j$ is the binary mapping: $\Omega \to \{true, false\}$ defined by the logical expression:*

$$a_j = \wedge_{i=1,p} e_i^j$$

*where $e_i^j$ is the event defined by $e_i^j = [y_i^j R_i^j V_i^j]$ and where $R_j = (R_1^j, \ldots, R_p^j)$ and $V_j = (V_1^j, \ldots, V_p^j)$.*

**Example 3** *a is an assertion defined by the triple $(a, R, V)$ where*
*$a = [size < 15] \wedge [color \in \{blue, white\}] \wedge [material \in \{crystal\}]$, $R = (<, \in, \in)$, and $V = ([15], \{blue, white\}, \{crystal\})$.*

When we are working on assertions defined not only on $\Omega$ but on different domains (i.e., on different observation spaces), we need to define the event taking care of the domains.

**Definition 4** *A **synthesis object** S is a symbolic object defined by the triple $S = (s, R, V)$ where s is the binary mapping: $\Omega^p \rightarrow \{true, false\}$ defined by the logical expression:*

$$s = \wedge_{j=1,q} a_j(\omega_j) = \wedge_{j=1,q} \wedge_{i=1,p} e_i^j(\omega_j) = \wedge_{j=1,q} \wedge_{i=1,p} [y_i^j(\omega_j) R_i^j V_i^j] = a_1(\omega_1) \wedge \ldots \wedge a_q(\omega_q)$$

*where $R = (R_i^j)_{i=1,p}^{j=1,q} = (R_1^1 \ldots R_1^q, R_2^1 \ldots R_2^q, R_p^1 \ldots R_p^q)$*
*and $V = (V_i^j)_{i=1,p}^{j=1,q} = (V_1^1 \ldots V_1^q, V_2^1 \ldots V_2^q, V_p^1 \ldots V_p^q)$.*

**Example 4** $S_1 = [size(vase) = 25] \wedge [color(flower) \in \{red\}]$.

**Definition 5** *Let $\Omega^p$ be sets of elementary objects. The **modal synthesis object** $S_{mod}$ is a synthesis object defined by the triple $S_{mod} = (s_m, R, V)$ where $s_m$ is a mapping $\Omega^p \rightarrow [0, 1]$ defined by the logical expression:*

$$s_m = a_1(\omega_1) \wedge \ldots \wedge a_q(\omega_q)$$

*where $a_i(\omega_i) = [y_i^j(\omega_i^j) \sim V_i^j] = [y_i^j(\omega_i^j) R_i^j V_i^j]$, $(R_i^j)_{i=1,p}^{j=1,q}$ is denoted by $\sim$,*
*$y_i(\omega_i) = (q_1^i(m_1^i), \ldots, q_{k_i}^i(m_{k_i}^i))$, $V_i = (p_1^i(m_1^i), \ldots, p_{k_i}^i(m_{k_i}^i))$ where $q_j^i, p_j^i \in [0, 1]$ and*
*$a_i(\omega_i) = \left(\sum_{j=1,k_i} q_j^i p_j^i\right) / k_i$ and $a(\omega) = \sum_{i=1,k} a_i(\omega_i)/k$.*

With this formula, it is possible to calculate the extent of the modal synthesis object and also, to allocate a new individual $\omega$ fitting the synthesis object.

**Example 5** $a(flower, vase) = [color(flower) \sim (0.5(pink), 0.4(yellow), 0.1(blue))]$
$\wedge [shape(vase) \sim (0.6(square), 0.2(round))]$.

If $y_1(flower_1) = (0.3, 0.2, 0.5)$ and $y_2(vase_2) = (0.4, 0.3)$, which means that $flower_1$ is 0.3 pink, 0.2 yellow and 0.5 blue and $vase_2$ is 0.4 square and 0.3 round then $a(flower1, vase2) = ((0.5 \times 0.3 + 0.4 \times 0.2 + 0.1 \times 0.5)/3 + (0.6 \times 0.4 + 0.2 \times 0.3)/2)/2$.

## 15.3 Generalization

In the machine learning domain, there are currently many methods used to generalize events (Michalski et al. 1983, 1986 [228, 229]), (Kodratoff 1986 [188]). We will recall here some of the main existing methods:

- taxonomy: This replaces each child by its closest ancestor.
  *Example:* by referring to the tree of taxonomic variables described in Figure 15.2, $a = [color \in \{clear\}]$ generalizes $a_l = [color \in \{white, yellow\}]$.

- "dropping condition rule": This deletes events that can take any value and keeps the simplest conjunctions of the most discriminating events.
  *Example:*
  $a_l = [color \in \{blue, green\}] \wedge [size \in \{large, medium\}]$ can be generalized by: $a = [color \in \{blue, green\}]$.

- "turning conjunction into disjunction rule": This groups values of the descriptors when descriptors are identical.
  *Example:*
  $a_l = [color \in \{blue\}] \wedge [size \in \{large, medium\}]$
  $a_2 = [color \in \{blue, green\}] \wedge [size \in \{small\}]$
  can be generalized by: $a = [color \in \{blue, green\} \wedge [size \in \{large, medium, small\}]$.

## 15.4   Background Knowledge

Background knowledge such as taxonomies and affinities can generalize, simplify, or combine different types or classes of symbolic objects. We will define all the background knowledge that we use in our algorithms as discussed below.

### 15.4.1   Taxonomies of Inheritance

A variable whose categories are ordered in a rooted hierarchical tree H is called a taxonomy. Such a variable will be called a taxonomic or tree-structured variable. An example is given in Figure 15.2.

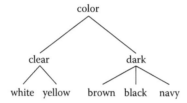

**FIGURE 15.2**
Example of a tree of the taxonomic variable color.

### 15.4.2   Affinities

We define affinity between two events. We denote $E_i$ as the set of events defined on $O_i$ and $E$ as the set of all events such that $E = \bigcup_i E_i$.

**Definition 6** *An affinity is a symmetric and positive mapping $aff : E \times E \to [0,1]$, which measures the degree of association or coherence between two events belonging respectively to two assertions defined on different domains. It is an expert knowledge.*

**Example 6** *We initially have elements of floral composition, vases, flowers and greenery, which are assertions and we seek to create synthesis objects represented by harmonious floral arrangements of different styles, i.e., composed of different kinds of flowers, vases, and greenery with good affinities between them.*
*Consider a set of affinities (aff) described as follows:*
$aff\left([color\,(flower) \in \{yellow\}]\,,[material\,(vase) \in \{crystal\}]\right) = 0.8,$
$aff\left([type\,(flower) \in \{iris\}]\,,[shape(vase) \in \{flared\}]\right) = 0.1$

### 15.4.3   Similarities

**Definition 7** *Let $a_i$ and $a_j$ be any two assertions in A, a **similarity measure** $r(a_i, a_j)$ is a function on $A \times A \to R^+$, positive and symmetric, which satisfies, for all $a_i, a_j \in A$, we denote $A = \{a_i\}_{i=1..n}$ the set of assertions.*

- $r(a_i, a_i) = r(a_j, a_j)$

- $r(a_i, a_i) \geq r(a_i, a_j)$

- $r(a_i, a_j) = r(a_j, a_i)$

- $r(a_i, a_j) \geq 0$

**Definition 8** *The Russell and Rao measure of similarity (Russel & Rao 1940 [279]) is defined on the $k^{th}$ variable between two symbolic objects $s_i, s_j$ as follows:*

$$r_{RR}(s_i^k, s_j^k) = \frac{n_{11}}{n_{11} + n_{10} + n_{01} + n_{00}} = \frac{n_{11}}{n}$$

*where $n_{11}$ is number of co-occurrences, the subscripts 0 and 1 represent respectively the absence and the presence of the occurrence of the category into the variable $k$, and $n$ is the total number of cases.*

**Example 7** *If we have:*
$a_1 = [color \in \{yellow, white\}]$
$a_2 = [color \in \{yellow\}]$
*The Russell and Rao measure of similarity between $a_1$ and $a_2$ is equal, for the variable color to:* $r_{RR}(a_1^{color}, a_2^{color}) = 1/2$.

## 15.5 The Problem

Once the measure of similarity between assertions of the same domain has been selected, the problem is the creation of synthesis objects from assertions based on affinities evaluated by experts. The input data are:

- a set of given assertions defined by events

- a similarity measure between events defined by the same variables (i.e., same domains)

- a set of affinity measure between events defined by different variables (i.e., different domains)

- the number of clusters.

The set of assertions may be such as: flowers defined by color, type, petal size ..., vases defined by their material, shape, size, color ..., greenery defined by price, season, ... and so on. The representation can be chosen among the set of vases. The output data that we are seeking to obtain are, first, assertion clusters and second, modal synthesis objects such as harmonious floral arrangements. To obtain this result, we apply an algorithm of dynamic clustering on symbolic objects, called SYNTHO that creates, after convergence, clusters of assertions that will be represented, after applying the algorithm of generalization, called GENOS, by a disjunction of modal synthesis objects. For example, we obtain, with the choice of representation $\{vase_1, vase_2, vase_6\}$ and after the clustering algorithm SYNTHO, a cluster $C_1 = (vase_1, flower_3, flower_4, greenery_2)$ consisting of a crystal vase, two types of flowers (rose, iris) and some greenery (eucalyptus), another cluster $C_2 = (vase_2, vase_3, flower_1, flower_4, flower_5)$ composed of two stoneware vases and three types of flowers (tulip, iris, marguerite), etc., following the number of clusters chosen initially. After the algorithm of generalization GENOS, we obtain modal synthesis objects corresponding to each cluster obtained and describe some harmonious floral arrangements such as:

$S_1 = [material(vase) \in \{crystal\}] \wedge [type(flower) \in \{rose, iris\}]$
$\wedge [color(flower) \sim \{0.75(clear), 0.25(violet)\}]$
$\wedge [size(flower) \sim \{0.6(large), 0.4(medium)\}]$
$\wedge [type(branch) \in \{eucalyptus\}].$

## 15.6 Dynamic Clustering Algorithm on Symbolic Objects: SYNTHO

The SYNTHO algorithm is a convergent algorithm of clustering (a type of dynamic clustering) (Diday 1971 [93]) on symbolic objects, which respects the four principles of symbolic data processing: fidelity to the data, predominance of knowledge, consistency, and interpretability (Diday 1987 [94]), (Kodratoff 1986 [188]). Indeed, it is guided by background knowledge represented by the affinities and taxonomies. Synthesis objects generated are more general than input assertions and synthesis objects obtained are easily explained, interpretable, understandable, and usable.

The data consist on the one hand of knowledge, represented by a list of "assertions" defined on different domains (observation spaces) and on the other hand, of additional knowledge provided by the expert, represented by taxonomies and defined affinities between "events" whose conjunction consists of "assertions", that we defined in Section 15.2.

### 15.6.1 Description of the Algorithm

Before describing the SYNTHO algorithm, it is necessary to provide the following three measures:

- a measure of similarity $r(a_i, a_j)$ or affinity $uff(a_i, a_j)$ between assertions, to measure the adequacy between two assertions.

- an index of aggregation $\Delta(a_i, P_j)$, which measures the similarity between an assertion and a cluster.

- a criterion $W(L, P)$, to optimize, measuring the adequacy between a partition $P$ of $A$ (a set of assertions) and its representation $L$.

  The purpose of the SYNTHO algorithm is to set up an optimization criterion $W(L, P)$, which is to be maximized through functions of assignment ($f$) and representation ($g$) that we define later.

We denote $A = \{a_i\}_{i=1..n}$ the set of given assertions to partition, $e_{a_i}^n$, as the $n^{th}$ event of assertion $a_i$ and $aff(e_{a_i}^n, e_{a_j}^m)$ as the affinity between two events $e_{a_i}^n$ and $e_{a_j}^m$, given by an expert.

We define the affinity $aff(a_i, a_j)$ between the assertions $a_i$ and $a_j$ such that:

$$aff(a_i, a_j) = \text{Max}[aff(e_{a_i}^n, e_{a_j}^m)].$$

In the case of assertions defined on the same variables, we use the Russell and Rao measure of similarity $r_{RR}(a_i, a_j)$ between two assertions, such as:

$$r_{RR}(a_i, a_j) = \frac{1}{n} \sum_{k=1}^{n} \frac{1}{p_k} \sum_{l=1}^{p_k} \delta_{ij}^l.$$

with $\delta_{ij}^l = \begin{cases} 0 & if \quad y_l^k(a_i) \neq y_l^k(a_j) \\ 1 & if \quad y_l^k(a_i) = y_l^k(a_j) \end{cases}$ where $y_l^k$ is the value of the $l^{th}$ category of the $k^{th}$ variable $y^k$ with $k \in \{1 \ldots n\}$ and $l \in \{1 \ldots p_k\}$, $p_k$ is the number of categories of the $k^{th}$ variable and $n$ is the number of variables of the assertions $a_i, a_j$.

In the case of assertions defined on different variables, we denote: $r'(a_i, a_j) = aff(a_i, a_j)$ but in the case of assertions defined on the same variables, we denote: $r'(a_i, a_j) = r_{RR}(a_i, a_j)$.

Let $P = \{P_j\}_{j=1 \ldots k}$ be a partition of $A$ into $k$ clusters. We define an index of aggregation $\Delta(a_i, P_j)$, between an assertion $a_i$ and the cluster $P_j$ such as:

$$\Delta(a_i, P_j) = \sum_j [r'(a_i, a_j), a_i \in L_i, a_j \in P_j]$$

where $r'$ is either an affinity $aff$ or a measure of similarity $r_{RR}$ depending on the fact that assertions are defined on the same variables or not

For $L_i$ being a representation:

$$\Delta(L_i, P_j) = \sum_j [r'(a_i, a_j), a_i \in L_i, a_j \in P_j].$$

We denote $L = \{L_1, \ldots, L_k\} \subset A$, the set of prototypes or the space of representation of a partition and define the criterion to be optimized $W(L, P)$, representing the improvement of the affinities and similarities between the prototype and the assertions of the cluster as:

$$W : L_k \times P_k \to R^+, \ W(L, P) = \sum_j [\Delta(L_j, P_j), j = 1 \ldots k].$$

We will demonstrate later that the value of this criterion increases with each iteration until it becomes stable.

The SYNTHO algorithm is iterative, proceeding by alternating between two steps and based on two functions:

- $f$, the assignment function, which associates a partition $P$ of $k$ classes with the k prototypes $L$

- $g$, the representation function, which associates with the $k$ prototypes $L$ a partition $P$ of $k$ classes.

  1. Given an initial set of $k$ prototypes $L = \{L_1, \ldots, L_k\}$, taken from the set of assertions $A = \{a_1, \ldots, a_n\}$, which may be specified randomly or according to criteria determined by the expert. The choice of $k$ will be based on the number of desired clusters.

  2. Assign each assertion $a_i \in A \backslash L$ to the cluster with the closest prototype $L_j \in L$ according to the criteria for assigning the assertion $a_i$ to the prototype $L_j$:
  $r'(a_i, L_j) = Max[r'(a_i, L_k), L_k \in L]$.
  It forms thus the cluster $P_j$ and we obtain a partition $P$ of $A$ with $P = \{P_1 \ldots P_j \ldots P_k\}$ of $A$.
  $f : L_k \to P_k$ where $f(L_j) = P_j$
  with $P_j = \{a_i \in A \backslash L, r'(a_i, L_j) > r'(a_i, L_m)$ with $L_m \in L \backslash L_j\}$.
  In the case where the criterion for assigning an assertion $a_p$ to two prototypes would be the same, we arbitrarily assign this assertion, $a_p$, to the prototype of smallest index. Thus, each $P_j$ is constituted by the relationship defined above.

3. Once the partition $P$ is given, we have to find all new prototypes:
$L = \{L_1 \ldots L_j \ldots L_k\} \subset A$ on $P = \{P_1 \ldots P_j \ldots P_k\}$. Each new prototype $L_j$ of $P_j$ will be determined as having the strongest representation criterion of cluster $P_j$ within $A, \Delta(L_j, P_j)$.

$g : P_k \to L_k$ where $g(P_j) = L_j$ with $L_j \in A, \Delta(L_j, P_j) = Max[\Delta(a_i, P_j), a_i \in A]$. The algorithm is deemed to have converged when the assignments no longer change.

### 15.6.2　Convergence of the Algorithm

The SYNTHO algorithm induces two sequences $U_n$ and $V_n$ defined as follows:
Let $U_n = W(L^n, P^n) \in R^+$, $V_n = (L^n, P^n) \in L_k \times P_k$ where $L^n = g(P^{(n-1)})$ and $P^n = f(L^n)$ where $U_n$ is a positive sequence, because it is a sum of positive values.

**Proposal 1** *The SYNTHO algorithm converges.*

**Proof 1** *We have to show that $f$ is increasing the value of the criterion $W$. In other words, we have to prove that:*

$$W(L, P) < W(L, f(L)), i.e., \sum_{j=1}^{k} \Delta(L_j^n, P_j^n) \leq \sum_{j=1}^{k} \Delta(L_j^n, P_j^{n+1})$$

*Which is equivalent to:*

$$\sum_{j=1}^{k} \sum_{a_j \in P_j^n} [r'(L_j^n, a_j^n)] \leq \sum_{j=1}^{k} \sum_{a_j \in P_j^n} [r'(L_j^n, a_j^{n+1})]$$

*Each assertion is assigned to the class $P_j^n$ if its affinity to $L_j^n$ is the highest among all prototypes. Therefore, the $a_i$ changing cluster increases the value of the criterion $W$. From which it results that:*

$$W(L^n, P^n) < W(L^n, P^{n+1}) \tag{15.1}$$

*We need now show that:*

$$W(L^n, P^{n+1}) \leq W(L^{n+1}, P^{n+1}) \tag{15.2}$$

*In other words, we have to prove that: $W(L, P) \leq W(g(P), P)$, which is equivalent to:*

$$\sum_{j=1}^{k} \Delta(L_j^n, P_j^n) \leq \sum_{j=1}^{k} \Delta(L_j^{n+1}, P_j^n)$$

*which is true by construction. For $j = 1, \ldots, k$, we have:*

$$\Delta(L_j^{n+1}, P_j^n) = Max[\Delta(a_i, P_j^n), a_i \in A]$$

*Hence, it results from (15.1) and (15.2) that: $W(L^n, P^n) \leq W(L^{n+1}, P^{n+1})$. Therefore, we have: $U_n \leq U_{n+1}$, which means that $U_n$ is an increasing sequence. As it is positive and bounded by the sum of all affinities for all the given events that can be combined, we can conclude that it is convergent.*

## 15.6.3   Building of Synthesis Objects

For each cluster $C_i$, we generate $S_i$, the synthesis object representing the cluster. The synthesis object $S_i$ will be characterized by the conjunction of assertions contained in the cluster $C_i$. The obtained synthesis object $S_i$ is a conjunction of assertions $a_i$, themselves a conjunction of elementary events. We represent $S_i$ as follows: $S_i$ is a mapping of $\Omega_1 \times \ldots \times \Omega_k \to \{true, false\}$ denoted by $S_i = \{\wedge a_i^j(\omega_j), a_i^j \in C_i\}$ where $\omega_j$ is a generic element of $\Omega_j$.

**Example 8 (of obtained cluster)** $C_1 = $ *(vase1, flower2, flower3) where*
$vase_1 = [color \in \{blue\}] \wedge [size \in \{large\}] \wedge [material \in \{crystal\}] \wedge [shape \in \{round\}]$
$flower_2 = [type \in \{rose\}] \wedge [color \in \{yellow, white\}] \wedge [size \in \{large\}] \wedge [stem \in \{thorny\}]$
$flower_3 = [type \in \{iris\}] \wedge [color \in \{yellow, violet\}] \wedge [size \in \{large, medium\}] \wedge [stem \in \{smooth\}]$.

The synthesis object $S$, not yet generalized, representing the obtained cluster $C_1$ will be:
$S_1 = [color(vase_1) \in \{blue\}] \wedge [size(vase_1) \in \{large\}] \wedge [material(vase_1) \in \{crystal\}]$
$\wedge [shape(vase_1) \in \{round\}] \wedge [type(flower_2) \in \{rose\}] \wedge [color(flower_2) \in \{yellow, white\}]$
$\wedge [size(flower_2) \in \{large\}] \wedge [stem(flower_2) \in \{thorny\}] \wedge [type(flower_3) \in \{iris\}]$
$\wedge [color(flower_3) \in \{yellow, violet\}] \wedge [size(flower_3) \in \{large, medium\}]$
$\wedge [stem(flower_3) \in \{smooth\}]$.

After three steps of generalization, that we will explain in Section 15.7, we obtained the following final modal synthesis object:
$S_{1_{mod}} = [material(vase) \in \{crystal\}] \wedge [color(flower) \in \{clear\}]$
$\qquad \wedge [type(flower) \sim \{0.5(rose), 0.3(tulip)\}] \wedge [size(flower) \sim \{0.6(large), 0.2(medium)\}]$.

## 15.7   Algorithm of Generalization: GENOS

We propose to apply a conceptual generalization algorithm on the clusters, obtained by the SYNTHO algorithm, to find an "explanation" of these clusters. Indeed, the purpose of acquisition of concepts is to find a characteristic description of a cluster of objects (Diday 1989 [96]).

Reminder: A symbolic object $s$ recognizes an event object $\omega \in \Omega$ if $\omega$ belongs to the extent of $s$.

GENOS generalization algorithm tries to generalize obtained clusters by a characteristic and discriminant recognition function. The generalization is accomplished using three methods (Michalski et al. 1983, 1986 [228, 229]), (Kodratoff 1986 [188]), that are applied in three steps. The first step of generalization is based on the "generalization tree climbing rule" also called "generalization rule by taxonomy", the second step of generalization is based on the "dropping condition rule", and the third step is based on the "turning conjunction into disjunction rule". The aim of this generalization is to transform clusters obtained by the algorithm SYNTHO into modal synthesis objects.

### 15.7.1    First Step of Generalization

This step is based on the fact that we work on structured data, i.e., there are some taxonomy trees defined on the data. We use the "generalization tree climbing rule": "Every child is replaced by its nearest ancestor".

**Example 9** *Example of a synthesis object obtained after the first step of generalization applied on $S_1$, referred as the defined taxonomy above in Section (15.4.1):*

$S_1' = [color(vase) \in \{blue\}] \wedge [size(vase) \in \{large\}] \wedge [material(vase) \in \{crystal\}]$
$\wedge \, [shape(vase) \in \{round\}] \wedge [type(flower) \in \{rose\}] \wedge [color(flower) \in \{clear\}]$
$\wedge \, [size(flower) \in \{large\}] \wedge [stem(flower) \in \{thorny\}] \wedge [type(flower) \in \{tulip\}]$
$\wedge \, [color(flower) \in \{clear\}] \wedge [size(flower) \in \{large, medium\}] \wedge [stem(flower) \in \{smooth\}].$

Indeed, in the example 8, we replace $flower_2$ and $flower_3$ by $flower$, $vase_1$ by $vase$, *yellow* and *white* by *clear*, through taxonomies initially defined on the data.

### 15.7.2    Second Step of Generalization

We want to obtain a description of a synthesis object, the simplest and the most discriminating description compared to other clusters. We use the following "dropping condition rule", which can give us a definition by intent of each cluster: We remove the events that may take any value and we keep the simplest conjunctions of the most discriminating events. The method is to choose the event to be kept depending on its degree of covering and its discriminating power. Thus, the choice of event that is retained will be obtained by searching for the most covering event (i.e., the one with the highest frequency in the cluster) and the most discriminating (i.e., the one with the lowest frequency of occurrence in the other clusters).

**Note 1** *We leave out $[y_i = O_i]$, and this omission is justified by the fact that the extent of $[y_j = V_j] \wedge [y_i = O_i]$ has the same extent as that for $[y_j = V_j]$. Note that if we want to consider the case where a variable has no meaning for an elementary object, we denote $[y_i = \emptyset]$ and we omit the events of the form $[y_i = O_i \cup \emptyset]$, which is justified by the fact that the extent of a symbolic object of the form $[y_j = V_j] \wedge [y_i = O_i \cup \emptyset]$, is identical to $[y_j = V_j]$. Note that $[y_j = V_j] \wedge [y_i = O_i \cup \emptyset]$ has, in its extent, elements such as $[y_i = \emptyset]$. Note also that for $[y_j = V_j]$, we can say that an elementary object such as $[y_j(\omega) = \emptyset]$ is not in its extent but we cannot say that it is false that it is there, but rather that it is not possible to say.*

We introduce two indicators to obtain the most discriminating and the most covering generalization of each cluster as follows:

1. An indicator of the degree of covering of each event to evaluate its power of generalization; we try to find the event that contains in its extent the maximum number of assertions (contained) in the cluster. We define the degree of covering $DR_{e_i}$ as follows:

$$DR_{e_i} = \frac{\text{Card}(Ext(e_i)_{C_i})}{\text{Card}(C_i)} \quad where$$

$\text{Card}(Ext(e_i)_{C_i}) =$ number of individuals in the extent of $e_i$ belonging to cluster $C_i$.

2. An indicator of the power of discrimination, which is an estimator of the tendency of a descriptor (or rather event in our formalization) to differentiate the individuals it

describes (Vignes 1991 [306]). We define the power of discrimination of an event by the ratio of discrimination $R_{(e_i)}$ (Hotu et al. 1988 [167]) as follows:

$$R_{(e_i)} = \frac{\text{Card}(Ext(e_i)_{C_i})}{\text{Card}(\Omega)}$$

The problem is to find a good trade-off between the homogeneity of the distribution of objects within the description set and the covering set that corresponds to the extent of the symbolic object, i.e., to find the best trade-off between the discrimination's level and the covering's level. The user can choose, at the time of generalization, a threshold of acceptable discrimination, which implies automatically the level of covering. If the user chooses a discrimination threshold of 1, he/she will obtain a generalization discriminating 100%, but necessarily less covering; if he/she accepts a discrimination threshold of 0.8, it will give less discriminating generalization (80%) but more covering. The user thus obtains a level of generalization corresponding to his choice of discrimination. It will depend on the data. Some data need a high discrimination level and lower covering level but some data need the opposite. We calculate the degree of covering and the ratio of discrimination of each event in each synthesis object representing a cluster. We keep in the description the event where the discrimination ratio is higher than the threshold. We will obtain the list of the most covering events, i.e., those that maximize the increase with the cardinality of the synthesis object's extent, ordered by their discriminating power compared to the other clusters. We retain the event with the maximum degree of covering and ratio of discrimination and, following the "dropping condition rule", we remove events belonging to the description of the retained events extent. We thus repeat this process on the conjunction of the remaining events and not yet generalized (computing of the degree and the ratio, maximum ratio event selection, deletion), until an acceptable discrimination threshold is reached. We obtain a disjunction of modal synthesis objects or Boolean objects in the special case where the threshold of discrimination is equal to 1. These objects, defined below, will be described by a conjunction of events, fitted with a mode with semantics corresponding here to a power of discrimination.

**Example 10** *If we apply the second step of generalization on $S_1'$ with a discrimination ratio equal to 1, we obtain:*
$S_{1_1}'' = [material(vase) \in \{crystal\}] \wedge [color(flower) \in \{clear\}].$
   *This means that this cluster contains all vases in crystal and all clear colored flowers. If we choose a discrimination ratio equal to 0.75, we could obtain*
$S_{1_{0.75}}'' = [material(vase) \in \{crystal\}] \wedge [color(flower) \in \{clear\}]$
$\qquad \wedge [type(flower) \sim 0.8(rose)] \wedge [type(flower) \sim 0.77(tulip)]$
$\qquad \wedge [size(flower) \sim 0.95(large)] \wedge [size(flower) \sim 0.81(medium)].$
   *This means that 100% of crystal vases, 100% clear colored flowers, 80% of roses, 77% of tulips, 95% of large flowers, and 81% of medium flowers are in this cluster.*

### 15.7.3   Modal Symbolic Objects and Modal Synthesis Objects

So far, the symbolic objects encountered satisfy a simple Boolean logic. Such logic is no longer applicable when we introduced into the representation of these objects some more complex knowledge or mode. Therefore, it is necessary to introduce modal symbolic objects (see definition 5) that need the use of so-called modal logics. The mode can express typically weights, probabilities or relative frequencies but also capacities, necessities, beliefs, or possibilities but here we just use the mode to express the discrimination ratio (see definition in Section 15.7.2). Here the mode $m_i(v)$ represents a frequency corresponding to the ratio of the cardinality's assertions with the modality in the cluster on the cardinality of the assertions with the modality in the total population, i.e.,

$$m_i(v) = \frac{\text{Card}(v/v \in C)}{\text{Card}(v/v \in \Omega)}$$

Note that the semantics of discrimination of $m_i(v)$ is not general but specific to our application domain.

**Example 11** *Let the modal synthesis object be:*
$S_{mod} = [color(flower) \sim \{0.8(pink), 0.4(yellow)\}]$
$\qquad \wedge [shape(vase) \sim \{0.6(square), 0.6(round)\}].$
*This means that 80% of pink flowers, 40% of yellow flowers, 60% square vases, and 60% round vases are in this cluster and it is described by the synthetic object $S_{mod}$.*

### 15.7.4 Third Step of Generalization

This step is based on the "turning conjunction into disjunction rule". It allows us to obtain the generalization of events composing the synthesis object by a union of descriptors of symbolic objects.

**Example 12** *>From the example 10, $S''_{1_{0.75}}$, we finally obtain:*
$S'''_{1_{0.75}} = [material(vase) \in \{crystal\}] \wedge [color(flower) \in \{clear\}]$
$\qquad \wedge [type(flower) \sim \{0.8(rose), 0.77(tulip)\}] \wedge [size(flower) \sim \{0.95(large), 0.81(medium)\}].$

---

## 15.8 Application: Advising the University of Algiers Students

The method is applied to the advising of the University of Algiers students, i.e., the choice of an academic sector for each Baccalaureate graduate, in order to improve the existing system. Currently, the graduates are advised on a given automatic and mandatory academic sector, based on criteria satisfying no one, neither teachers nor students. The criteria are, for instance, criteria of vacancies per course, but without real relevance to the student's level or his/her affinities. With our method, we want to improve the system by making it more flexible, efficient, and closer to the affinities of each student. What is the academic sector that suits the student best, according to a given Baccalaureate graduate's profile, by the main Baccalaureate marks and preferences? That is the question we wish to answer with our "synthesis object method". Our method is based on the concept of affinity between a student and an academic course. This affinity is obtained from some "background knowledge" and "expert knowledge", as determined by the domain expert. It can be given as a positive number in $\{0, 1\}$ or as a calculation of scale. Here, the expert knowledge is the scale specific to each academic sector, depending on Baccalaureate specialization, marks, and grades. From the affinities and the use of the SYNTHO algorithm, we obtain a classification where the clusters obtained contain academic sectors and students with a maximum affinity with them. These clusters are expressed in the form of synthesis objects, which are, after a phase of generalization, expressed in the form of modal synthesis objects.

## 15.8.1   Input Data

The input data are described in two groups of assertions described by different sets of variables. On the one hand, we have all academic sectors, and on the other hand, we have all students described by their specialization, their distinction, and their Baccalaureate grade obtained in mathematics, physics, and biology.

*Set of students:* In our database, we have 332 students enrolled in 1984, 1985, 1986, and 1987 in biology and sciences academic disciplines. All scores are out of 20. A student will be described as follows for example:
$e_1 = [bac \in \{s\}] \wedge [dist \in \{p\}] \wedge [math = 11.5] \wedge [phys = 9.5] \wedge [bio = 14]$.

This is a student who earns the scientific Baccalaureate, having distinction as a pass (i.e., he had an average score) whose marks are 11.5 in mathematics, 9.5 in physics, and 14 in biology.

*Set of academics sectors:* There are 7 academic sectors.
The different academic sectors are:
f1 = [academic sector = sciences]
f2 = [academic sector = engineering]
f3 = [academic sector = computer]
f4 = [academic sector = electronic]
f5 = [academic sector = electrical engineering]
f6 = [academic sector = biology]
f7 = [academic sector = senior technician].

*Set of affinities:* There are 12 affinities of expert knowledge, developed in collaboration with the expert to find the academic sector that best suits each student. Each affinity is an average of weighted marks, each calculation being different depending on the academic sector and the Baccalaureate's specialization.

*For example,* affinity calculated between a Baccalaureate graduate with a mathematical specialization and academic course biology will correspond to the following calculation:
  *Affinity (academic sector $\in \{biology\}$, $bac \in \{m\}$) $= ((bio * 2.5) + phys + math) /8$.*
Then, depending on the maximum affinity, a student will be assigned to a particular academic sector.

We restrict the choice of the representations to be only among *academic sectors*. The algorithm starts with five *academic sectors* and the representations remain among the *academic sectors*.

## 15.8.2   Results

In analyzing the results, we are looking for two specific objectives:

1. To verify at the level of the classification algorithm on symbolic objects, that in the majority of cases, students assigned to an academic sector are found in their original academic sector. Remember that our dataset includes students already at the university. It is also interesting to check, in cases where a student is oriented towards another sector, if it is justified and improves his/her choices.

2. To check, at the level of the generalization algorithm, that the obtained modal synthesis objects give a satisfactory typology of students for each academic sector.

For the classification phase, in the SYNTIIO algorithm, we get five clusters: one cluster containing unclassifiable students and four clusters containing respectively students and the corresponding academic sectors. We verified that in 90% of cases, students found themselves in their original academic choice. In less than 10% of cases are students assigned to other sectors, and we realize that this declassification is fully justified in relation to our objectives. For example, among the dataset consisting of 70 students enrolled in biology in 1987, we had 63 students who are being reassigned to biology. Among the 7 not reassigned (10%), we have:

- Two are assigned to {sciences, engineering} because their biology level is low and their mathematics level is high.

- One is assigned to {computer, electronics, electrical engineering} because his biology level is low but by contrast, his mathematics and physics level is high.

- Two are assigned to {senior technician} as their level is weak overall.

- Two are assigned to {unclassifiable} cluster having a "bc" Baccalaureate specialization not listed initially.

We can conclude that our first objective has been reached. Indeed, the classification algorithm SYNTHO based on the affinities gives us not only correct and interesting results but also significantly improves the previous orientation.

For the generalization phase of the GENOS algorithm, by allowing the approach of the profile of students by academic sector, the results are composed of five modal synthesis objects corresponding respectively to the five clusters obtained previously. The description of each modal synthesis object can be multiple, depending on the threshold chosen by the user or the expert, and according to this threshold, we obtain a generalized description of a modal synthesis object more or less discriminating, as illustrated by the following results.

By choosing a discrimination threshold equal to 0.5, five profiles of students by academic sectors were highlighted as, for example, in the following description.

$Cluster_{3_{0.5}} = [academicsector \in \{computer, electronic, electricalengineering\}] \wedge [distinction \in \{verygood\}] \wedge [phys \sim \{0.6(15.5), 0.5(16.5, 16)\}]$.

$Cluster_3$ corresponds to "computer, electronic, and electrical engineering" academic sectors and is composed of all students who obtained the distinction "very good" (discrimination mode is equal to 1), 60% of students who had 15.5 in physics and 50% of students who had 16 or 16.5 in physics. If we chose a discrimination threshold equal to 1, we obtain for $cluster_3$:

$Cluster_{3_1} = [academicsector \in \{computer, electronic, electricalengineering\}] \wedge [distinction \in \{verygood\}]$.

We could conclude from the previous results that each obtained modal synthesis object corresponds to the choice of the expert concerning the profiles of students associated with each academic sector. The generalization phase corresponds very well to the second objective, which is to obtain a satisfactory students' profile by academic sector. It is then possible to allocate a new student to the synthesis object that it fits the best, as explained in Section 15.2.

## 15.9    Conclusion

The new approach of Symbolic Data Analysis has been introduced in this chapter. Indeed, the integration between symbolic and numeric approaches is to combine the advantages of each of them: ease of reporting and understanding of symbolic approaches, and accuracy of numerical approaches.

More specifically, this aspect of the classification on symbolic objects used the method of dynamic clustering, adapted to these objects. The concepts of classification on symbolic objects, affinities, and generalization have been introduced and implemented by the SYN-THO and GENOS algorithms. This work can be useful for handling complex knowledge such as assertions with taxonomies and affinities, or modal synthesis objects.

We conclude by giving an overview of extensions that we propose to be realized in the future.

- About this application, to complete the affinities with the wishes of the students, by taking into account their preferences, which our Knowledge Database does not include at present and we are therefore limited by the affinities of levels.

- More generally, the introduction of constraints as objects rules replacing or supplementing some affinities to reduce the amount of background knowledge, and, the creation of an interactive system allowing changes to the background knowledge and assignment for a new student.

# 16

## Functional Data Analysis: An Interdisciplinary Statistical Topic

**Laurent Delsol, Frédéric Ferraty, and Adela Martínez Calvo**

## CONTENTS

## 16.1  Introduction

The progress of electronic devices like sensors allow us to collect data over finer and finer grids. These high-tech data introduce some continuum that leads the statisticians to consider statistical units as functions, surfaces, or more generally as a mathematical object. The generic terminology for this new kind of data is Functional Data (FD) and the statistical methodologies using FD are usually called Functional Data Analysis (FDA). FD appear in numerous fields of sciences like biology, chemometrics, econometrics, geophysics, medicine, etc. The usefulness of FDA as a decision-making tool in major questions like health, environment, etc., makes it really attractive for the statistician community and this is certainly the main motor of its fast development. In that sense, FDA is really an interdisciplinary and modern statistical topic. Several books promoted the practical impact of FDA in various fields of sciences ((Ramsay & Silverman 2002 [266]), (Ramsay & Silverman 2005 [267]), (Ferraty & Vieu 2006 [126]), and (Ferraty & Romain 2011 [124])).

    The aim of this contribution is to provide a survey voluntarily oriented towards the interactions between FDA and various fields of sciences whereas methodological aspects may be found in (González Manteiga & Vieu 2011 [147]). Because of the numerous works done in this area, it is hard to present an exhaustive bibliography. So, we decided to focus on the most recent advances instead of a general survey. One of the main advantages of this choice is to propose a snapshot in a research area where progress is extremely fast. This contribution is split into two sections. The first one (Section 16.2) gives several examples of FD and describes the various statistical problematics involving FD. Section 16.3 highlights the wide scope of scientific fields concerned with FDA by describing representative applied contributions in the most recent literature.

## 16.2    FDA Background

### 16.2.1    What Is a Functional Dataset?

The main idea in FDA is to consider that the statistical sample is composed of $n$ objects $X_1, \ldots, X_n$ belonging to some infinite-dimensional space. In most situations each $X_i$ can be a curve observed for the $i$th statistical unit

$$X_i = \{X_i(t), a \leq t \leq b\}, \ i = 1, \ldots, n,$$

but some situations may involve more complicated functional elements (such as images for instance). Section 16.2.2 will briefly present three different curves datasets, selected in order to highlight both the wide scope of possible applications and the variety of statistical questions arisen. In some situations, functional data need some preliminary treatment before being statistically usable, and Section 16.2.3 discusses the main recent bibliography in this direction. Then Section 16.2.4 points out general problematics that one could have to deal with in FDA.

### 16.2.2    Some Examples of Functional Datasets

*Example 1: Speech recognition data.* The first example comes from speech recognition sciences. The dataset contains 2000 curves that are log-periodograms of recorded voices corresponding to five different phonemes (sh, iy, dcl, a, ao). Each recording is sampled at 150 different frequences (i.e., 150 measurements), meaning that the data are discretized curves on a equispaced grid of size 150. Figure 16.1 presents a sample of 50 among these curves (10 for each phoneme). These data have been widely treated in the literature and the reader may find more details on this curves dataset (and instructions for getting them) in (Ferraty & Vieu 2006 [126]).

*Example 2: Pork data.* The second example comes from chemometrics. It turns out that spectrometric analysis is quite often used as a fast and low cost tool in quantitative chemical sciences, and the reader may find many spectrometric curves datasets both in the statistical literature and on the web. In fact, spectrometry is one of the first scientific fields that reached the attention of the statistical community because of its functional nature. The example that we have selected comes from the food industry. It was chosen because it became a benchmark dataset in the scientific community to check the validity of new statistical methods developed in FDA. This dataset provides 215 spectrometric curves obtained from 215 pieces of finely chopped pork meat. Each spectrometric curve is obtained

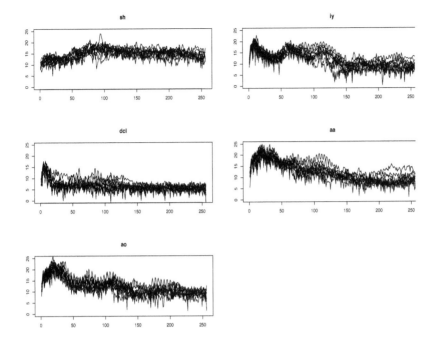

**FIGURE 16.1**
Phoneme data: samples of 50 log-periodograms.

from a grid of 100 different wavelengths. These data have been widely studied in the literature, and the reader may look at (Ferraty & Vieu 2006 [126]) for further presentations as well as for an indication on how to download them. The left panel of Figure 16.2 presents a sample of spectrometric curves; the right panel of Figure 16.2 displays the corresponding second derivatives that are, from chemometrical knowledge, more informative than the curves themselves.

*Example 3: El Niño data.* The final example comes from climatology. The data are monthly temperatures of sea surface observed in the period 1950–2006. This is a classical time series that is presented in Figure 16.3 (left panel).

A functional approach to this time series consists of cutting it in various paths in such a way that the data can be seen as a curves dataset. Here, the 684 monthly observed values are taken as 57 yearly curves, each being discretized at 12 points. This functional data representation of the time series is shown in Figure 16.3 (right panel). We refer to (Ferraty & Vieu 2006 [126]) for more details on El Niño applications, including extensive presentation of data, guidelines for downloading them, and statistical treatments by FDA methods.

### 16.2.3 Preprocessing Functional Data

In practice one does not have at hand the continuous curves $X_1, \ldots, X_n$ but only discretized versions of these curves

$$X_i(t_{i,j}), \ i = 1, \ldots, n, \ j = 1, \ldots, n_i.$$

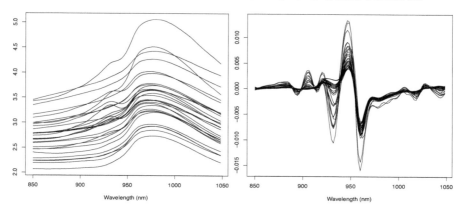

**FIGURE 16.2**

Pork data: sample of spectrometric curves (left panel) and corresponding second derivatives (right panel).

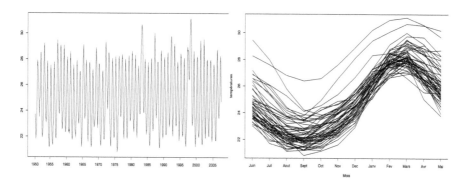

**FIGURE 16.3**

El Niño data: time series of monthly temperature 1950–2006 (left panel) and corresponding 57 yearly curves for monthly temperature (right panel).

Therefore, the first natural preprocessing of functional data consists of using some standard nonparametric smoothing procedure for transforming each discretized curves $X_i(t_{i,j}), j = 1, \ldots, n_i$ in the continuous data $X_i$. Any standard nonparametric method can be used for that, including kernel, splines, wavelets, or more sophisticated ones according to the shape of the curves (smooth curves, rough paths, etc.). Of course, this presmoothing treatment is much less important in situations when the discretization grid is fine (as in Figures 16.2 and 16.1) than for "sparsely" observed curves (as in the right panel of Figure 16.3). The presmoothing treatment is also very helpful in situations of missing data (i.e., for some $t_{i,j}$ the value $X_i(t_{i,j})$ is not observed), or more generally with unbalanced curves (i.e., $\exists (i, i'), n_i \neq n_{i'}$). Another situation when presmoothing is important is when each discretized data point $X_i(t_{i,j})$ involves some error of measurement (see Crambes et al. [77]).

Another important preprocessing of functional data is the so-called curve registration. This is used in situations when one wishes to suppress some artificial effect (like vertical or horizontal shift for instance). The spectrometric example (see left panel of Figure 16.2) is typically concerned with this point, since there is a vertical shift in the curves (known in chemometric sciences as a calibration effect), which has definitively no link with the chemical structure to be studied. One way for preprocessing this kind of spectrometic data is to consider data derivatives (see right panel of Figure 16.2). More generally, statistical methods have been developed to estimate some common feature of the curves (see Park et al. 2009 [251]) and/or to align some specific benchmarks (see Liu & Yang 2009 [213]). Another important problem in curve preprocessing is the detection and the estimation of some transformation aligning the whole collection of curves (see Kneip et al. 2000 [187]). All these methods include those known in the literature as curve registration and warping.

### 16.2.4    Some General Statistical Problematics Involving Functional Data

With a functional dataset, all the statistical questions occuring with standard multivariate data may also appear. Basically, these questions can be divided into two blocks. The first category concerns the exploration of the structure of some functional dataset, while the other one is linked with explanatory problems aiming to use some functional dataset for predicting some extra response variable.

The first category includes standard questions like the following ones. Can we define/estimate some centrality parameter (mean curve? modal curve? median curve?) for exploring the various yearly temperature curves depicted in the right panel of Figure 16.3? Can we observe/construct clusters in the various spectrometric curves depicted in Figure 16.2?

The second category of problems involves standard prediction questions. Can we use the spectrometric curves depicted in Figure 16.2 to predict, by means of some regression technique for example, some chemical features of the pork meat (fatness, protein, ...)? Can we use the speech recognition data depicted in Figure 16.1 for assigning a new log-periodogram, by means for instance of some discrimination technique, to some specific phoneme? Can we use the yearly climatic datasets depicted in the right panel of Figure 16.3 to predict, by means of some time series analysis approach, some characteristic of the future? All these interesting questions are discussed in (González Manteiga & Vieu 2011 [147]).

## 16.3    FDA: A Useful Statistical Tool in Numerous Fields of Application

The aim of this section is to present recent applied studies involving functional data. Most of these contributions have been classified according to the scientific field that is concerned (see Sections 16.3.1–16.3.7). A few other ones (see Section 16.3.8) are not directly associated with some specific science but appear indirectly through statistical analysis of other data. General contributions devoted to applied issues in FDA involve (Ramsay & Silverman 2002 [266]) and (Ferraty & Vieu 2006 [126]).

### 16.3.1    Chemometrics

As discussed before in Section 16.2.2 spectrometry is one of the most popular fields for which FDA methods are known to be of interest. In addition to the references concerning the benchmark pork dataset discussed before in Example 2, many other spectrometric curves have been analyzed by FDA methods (see for instance (Ferraty et al. 2010 [121])).

### 16.3.2    Medicine

Maybe one of the most common situations involving functional data analysis is growth curves. Recent advances on growth curves analysis include (Liu & Yang 2009 [213]). Other medical problems treated by FDA techniques concern for instance cancerology (Baladan-dayuthapani et al. 2008 [11]), placebo effects (Tarpey et al. 2003 [291]), cardiology (Di et al. 2009 [92]), ophthalmology (Locantore et al. 1999 [214]), genetics (Parker & Wen 2009 [252]), and mortality (Chiou & Müller 2009 [70]). FDA methods have also been applied for studying various human motions (knee, lips, hip, ...) by (López-Pintado & Romo 2007 [215]) or for analyzing human tactile perception (see (Spitzner et al. 2003 [287])).

### 16.3.3    Biostatistics

Biostatistics is a quite active current scientific field, and FDA methods find there a wide scope of possible applications (see [160] for a recent handbook on biostatistics including several contributions of the FDA community). Recent applications of FDA methods in genetics involve for instance the work of (Tang & Müller 2009 [289]). FDA methods have been also successfully applied in animal biology problems by (Chiou & Müller 2007 [69]).

### 16.3.4    Environmetrics and Climatology

Environmetrical sciences and climatology involve now more and more often data of a functional nature. In addition to the work (Ferraty et al. 2005 [123]) on the El Niño phenomenon discussed before in Section 16.2.2, other climatological problems studied by FDA techniques can be found in (Hall & Vial 2006 [152]). In environmetrics FDA can be used in various settings and one can note the contribution (Fernández de Castro & González Manteiga 2008 [119]) on the air pollution issue, (Nérini & Ghattas 2007 [241]) on water quality control and (Hlubinka & Prchal 2007 [166]) on atmospheric radioactivity. See also (Escabias et al. 2005 [113]) for general functional considerations on environmental data.

### 16.3.5    Signal Processing

A remote sensing satellite is a mechanism that produces large sets of functional data (radar curves, images, ...), and FDA finds there an interesting field of application. While in (Ramsay & Silverman 2002 [266]) some meteorological curves are analyzed, the paper (Cardot & Sarda 2006 [64]) studies satellite data for vegetations evolution. Other signal processing applications of FDA involve sound signal recognition (see Ferraty & Vieu 2006 [126] for speech recognition data or Hall et al. 2001 [151] for radar waveforms). Image processing is also a good example where FDA is useful (see for instance Jiang et al. 2009 [181]).

### 16.3.6    Econometrics

The economy is also a field for which people are quite often confronted with curves data, and the recent advances on FDA are of particular interest. Most applications are related to continuous time series that are analyzed in a functional fashion by means of the splitting preprocessing techniques as described under Example 3 in Section 16.2.2 above. Works in this field include (Wang et al. 2008 [310]) and (Liu & Müller 2009 [210]). Other applications dealing with sparsely observed stochastic processes can be found in (Peng & Müller 2008 [254]) and (Liu & Müller 2009 [210]).

### 16.3.7 Other Scientific Fields Involving Functional Data

Many other scientific fields are concerned with functional data, such as for instance graphology (Ramsay & Silverman 2002 [266])), geology (Nérini & Ghattas 2007 [241]), linguistics (Aston et al. 2010 [9]), or demography (Hyndman & Ullah 2007 [172]). The book by (Ramsay & Silverman 2002 [266]) presents also a wide scope of applications, including criminology, paleoanthropology, or education sciences.

### 16.3.8 Miscellaneous

To finish this section on possible applications of FDA methods, it is worth mentioning that functional data may appear indirectly through statistical treatments of standard multivariate data. For instance there may be situations where, for the $i$th statistical unit ($i = 1, \ldots, n$), one observes a $n_i$-sample $Z_{i,1}, \ldots, Z_{i,n_i}$ drawn by some multivariate variable $Z_i$. One may be interested in functional distributional features (for instance density $f_i$, cumulative distribution function $F_i$, etc.) of $Z_i$. From the $n_i$-sample $Z_{i,1}, \ldots, Z_{i,n_i}$, it is easy to build a $n$-sample of functional data by estimating these distributional features (i.e., $\widehat{f}_1, \ldots, \widehat{f}_n$, or $\widehat{F}_1, \ldots, \widehat{F}_n$, or any other estimated functional features). Works in this direction involve the analysis of a collection of density functions (Delicado 2007 [84]), a set of regression functions (Heckman & Zamar 2000 [164]), or a set of some probabilistic functional characteristic (Rossi et al. 2002 [275]).

## 16.4 Conclusions

This overview shows the obvious interaction between FDA and numerous fields of sciences. The progress of technologies will certainly reinforce the use of FDA to handle such high-tech datasets. It is worth noting that even if most of the methodological advances on FDA are, in theory, available for any kind of infinite-dimensional observed variables (curves, images, arrays, etc.), most of the case studies carried out by recent FDA techniques are devoted to curve data analysis. Undoubtedly, a very attractive open challenge for statisticians is to be able to solve the numerous applied issues linked with the application of FDA methodologies to functional objects that are more complicated than standard curves. Among them, image data could be considered with great attention. Definitely, FDA is already a promising statistical research area that should be intensively developed in the near future.

## Acknowledgments

The authors wish to thank the participants in the working group STAPH in Toulouse (France). The activities of this group on FDA and on various other features of Statistics in infinite-dimensional situations have been a continuous support and have hence contributed greatly to this wide bibliographical study. Special thanks to Wenceslao Gonzalez-Manteiga and Philippe Vieu who have actively collaborated in this survey work linked with their own contribution to this book (see González Manteiga & Vieu 2011 [147]). The research of the third author was supported by the national project MTM2008-03010 and the regional projects PGIDIT07PXIB207031PR and 10MDS207015PR.

# 17

# Methodological Richness of Functional Data Analysis

Wenceslao Gonzàlez Manteiga and Philippe Vieu

## CONTENTS

## 17.1   Introduction

Functional datasets become each day more and more present in many fields of applied sciences (see the contribution by (Delsol et al. 2011 [85]) in this volume) and the statistical community has started to develop specific tools for analyzing them. This active field of modern statistics is known as Functional Data Analysis (FDA). Even if first works in the literature trace back to earlier times, FDA became really popular at the end of the nineties with the book by Ramsay and Silverman (see (Ramsay & Silverman 2005 [267])). This great interest in FDA and the wide scope of its various facets can be seen through recent monographs or handbooks (Ramsay & Silverman 2005 [267], (Ferraty & Vieu 2006 [126]), and (Ferraty & Romain 2011 [124])) and through various special issues in top level statistical journals as *Statistica Sinica* (2004, vol. 14), *Computational Statistics and Data Analysis* (2007, vol. 51), *Computational Statistics* (2007, vol. 22), or *Journal of Multivariate Analysis*, (2010, vol. 101). The aim of this contribution is to present the state of art in statistical methodology for FDA. Given the increasing interest in the literature in this topic, this bibliographical discussion cannot be exhaustive. So, it will be concentrated on the precursor works and on the most recent advances. Section 17.2 gives special attention to the various spectral techniques that are widely used as preliminary benchmark methods in FDA. Then Section 17.3 presents the recent developments for exploratory analysis of functional data, while Section 17.4 discusses contributions in problems where functional data are used as explanatory variables. Section 17.5 discusses miscellaneous issues and Section 17.6 sums up the major issues remaining to be considered.

## 17.2 Spectral Analysis: Benchmark Methods in FDA

Starting with the earliest ideas involving functional variables, factor-based analyses turn out to be key benchmark tools for preliminary analysis of functional data. Spectral decomposition of some functional operator is in the heart of factor-based analyses. The operator on which the spectral analysis is performed may differ according to the problem one has to answer and also to the kind of statistical methodology one wishes to develop. Given the high interest of these methods, it is not possible to present an exhaustive overview of the literature. This section is devoted to the presentation of the state of art in factor-based analysis of functional data through a selected set of references, chosen mainly among the most recent ones.

### 17.2.1 Functional Principal Component Analysis

Functional principal component analysis (FPCA) consists mainly of projecting a sample of functional data $X_1, \ldots, X_n$ on a data-driven basis by means of the eigen-elements of their covariance operator. The functional principal components obtained by this way, or at least a small subset among them, are much more easy to deal with than the functional data themselves, both from an exploratory point of view and for further methodological problems. This question received some attention at the beginning of FDA but the literature increased a lot since the nineties. The book by (Ramsay & Silverman 2005 [267]) provides a general view on FPCA. Some works include extensions of PCA for bi-functional data (Spitzner et al. 2003 [287]). Methods and results concerning specifically the estimation of the covariance operator include (Rice & Silverman 1991 [270]), (Hall & Vial 2006 [152]), and references therein. The reader may find in (Park et al. 2009 [251]) a recent development of the concept of structural components that are more informative and easier to interpret than the principal components.

### 17.2.2 Other Spectral Analysis

Canonical analysis concerns the case when the statistical sample is composed of two observed variables $(X_1, Y_1, \ldots, X_n, Y_n)$ and when the question is to describe the links between $X$ and $Y$. This has been widely studied for multivariate variables and extensions have been proposed for functional ones. Recent advances in Functional Canonical Analysis can be found in (Kupresanin et al. 2010 [193]). A wide scope of the literature is presented along with some comparative study in (He et al. 2004 [163]). A general discussion on Functional Canonical Analysis can be found in chapter 11 of (Ramsay & Silverman 2005 [267]).

Functional Linear Discrimination Analysis (FLDA) is linked with canonical analysis since the data are of the form $(X_1, Y_1, \ldots, X_n, Y_n)$ but with $Y$ being valued in a finite set. The interest in using penalization ideas is discussed in (Hastie et al. 1995 [161]) and (Ramsay & Silverman 2005 [267]). The paper by (James & Hastie 2001 [178]) combines Functional

Linear discrimination with a preliminary dimensionality reduction approach, while (Preda et al. 2007 [261]) combines it with a functional version of partial least squares ideas.

When a sample of curves is collected, there exist some situations where the curves may be observed at few irregularly spaced measurements (i.e., the measurements may differ from one curve to another one), and a specific spectral analysis, called principal analysis by conditional expectation (PACE), has been developed for that (see Müller 2009 [232]) for the most recent advances).

## 17.3 Exploratory Methods in FDA

In addition to the spectral analysis discussed just before, various other methodologies have been introduced for exploring some functional variable $X$ through a sample dataset $X_1, \ldots, X_n$. Basically, one can split these contributions into two categories. The first one (see Section 17.3.1) focuses on various distributional descriptors (mainly dealing with centrality notions) that can be introduced for exploring functional data. The second category (see Section 17.3.2) deals with unsupervised classification issues for which the main goal is to build/identify existing clusters in the sample. Of course, this splitting into two categories is rather arbitrary, since in most cases distributional descriptors are important tools for depicting clusters. Note that defining distributional descriptors is not an easy task, since it requires the use of a density notion in functional spaces where a reference measure does not exist as the Lebesgue measure does in finite-dimensional space.

### 17.3.1 About Descriptors of Distribution in FDA

*About mean curve.* As for multivariate data, the first natural idea for looking at the centrality of some functional dataset is the notion of mean. For curve datasets, the naive idea of just computing for each $t$ the mean of the values $X_i(t)$ is of course unsatisfactory, but it has been adapted to a functional setting in various ways. Robust versions of the mean curve can be constructed by considering trimmed means, the main issue being how to define *central notion* with functional data. This is most often done by using a functional version of depth notion (see (López-Pintado & Romo 2009 [216]) for the most recent advances). See however (Cuesta & Fraiman 2007 [79]) for a direct approach not based on depth notion.

*Other centrality descriptors.* The first alternative to mean curve is the notion of median. A wide bibliographical discussion is provided in chapter 9 of (Ferraty & Vieu 2006 [126]). Of course, the functional depth notion discussed before can be directly used to define and estimate other notions of median (see again (López-Pintado & Romo 2009 [216]) and references therein). Another alternative to mean curve is the notion of modal curve as proposed in chapter 9 of (Ferraty & Vieu 2006 [126]).

*Other distributional descriptors.* In addition to these various centrality descriptors, a few other ones have been introduced in FDA. For instance, still by means of functional depth consideration, the notion of quantile can be extended to functional variables (see again (López-Pintado & Romo 2009 [216]) and references therein). Finally, other descriptors can be defined by considering the functional principal components (as described above in Section 17.2) rather than the data themselves, and in this case the challenge is to estimate the covariance operator or some benchmark of its components (see (Hall & Vial 2006 [152]) and references therein).

### 17.3.2   Functional Data Classification

Classification is an important topic in FDA when one wishes to detect some clusters in the data, and the centrality notions discussed above are important tools for reaching this goal. Chapter 9 in (Ferraty & Vieu 2006 [126]) presents a general classification algorithm usable with any centrality descriptors. Another way for doing functional classification consists of adapting suitably the k-means notion to functional data (see (Cuesta & Fraiman 2007 [79]) and references therein). The contribution by (James & Sugar 2003 [179]) is specifically adapted to sparsely observed curves. The paper by (Nérini & Ghattas 2007 [241]) is based on regression trees.

## 17.4   Explanatory Methods in FDA

Explanatory models and methods are linked with problems in which the data are of the form $(X_1, Y_1, \ldots, X_n, Y_n)$ and for which the question is to focus the attention on the links between $X$ and $Y$. The aim of this section is to present the state of art in this field when one (at least) among the variables $X$ and $Y$ is functional. Special attention is paid in Section 17.4.1 to the case when the response $Y$ is real valued and $X$ is functional, while other situations are discussed in Section 17.4.2. Functional discrimination (that is when $Y$ takes values in a finite set) will be discussed specifically in Section 17.4.3. Section 17.4.4 is concerned with predicting future values of a time series.

### 17.4.1   Predicting Scalar Response from Functional Covariate

*Prediction by conditional mean.* One of the most popular families of prediction techniques is based on conditional expectation, and consists of introducing a regression model $Y = r(X) + error$. The nature of the model depends on the kind of conditions assumed on the operator $r$ to be estimated and the statistical techniques to be introduced will be directly linked to the nature of the model. A classification of regression models is given in chapter 1 of (Ferraty & Romain 2011 [124]) based on three wide families: parametric, nonparametric, and semiparametric models.

The literature on parametric models is based on the functional linear model. Various estimation techniques have been proposed in this context, as for instance, those based on ridge regression ideas or total least square ideas (Cardot et al. 2007 [63]) and those based on presmoothing ideas (Ferraty et al. 2011 [120]). Various theoretical advances have been provided more recently, with the main goal being to make precise the rates of convergence of the different functional linear regressors: see (Crambes et al. 2009 [77]) for the most recent developments and references therein. Other discussions on functional linear regression models can be found in (Ramsay & Silverman 2005 [267]) and a complete recent survey will be carried out in chapter 2 of (Ferraty & Romain 2011 [124]). The generalized functional linear model has been studied (see (Li et al. 2010 [207]) for the most recent advances). The special case of functional logit model has been investigated in (Escabias et al. 2004 [112])).

Nonparametric models for functional regression has been popularized by (Ferraty & Vieu 2006 [126]) (see also chapter 4 of (Ferraty & Romain 2011 [124]) for a more recent survey). This field started with methods based on Nadaraya-Watson kernel estimators: see (Ferraty et al. 2010 [125]) and (Ferraty et al. 2010 [122]) for the most recent advances, while (Benhenni et al. 2007 [16]) investigates applied issues linked to automatic smoothing parameter choice.

Alternative nonparametric methods include k-NN estimation (Burba et al. 2009 [60]) and local linear estimation (see (Barrientos Marin et al. 2010 [14]) and references therein).

The wide literature devoted to dimensionality reduction in large (but finite) dimensional problems is starting to be adapted to functional data, and recently the literature in functional regression has developed a few semiparametric models with the aim to balance the trade-off between linear and nonparametric models (see chapter 4 of (Ferraty & Romain 2011 [124]) for a recent survey). This includes functional adaptation of Slice Inverse Regression (see (Hsing & Ren 2009 [169]) and references therein), of Additive Model (Ferraty & Vieu 2009 [127]), of Single Index Model (see (Ait Saidi et al. 2008 [3]) and references therein), and of Partial Linear Model (see (Aneiros Perez & Vieu 2008 [6]) for the most recent advance). Semiparametric functional data analysis is both underdeveloped and promising, and it will certainly be the object of much attention in the future.

*Prediction based on other characteristics of the conditional distribution.* As in multivariate data analysis, there exist other techniques than those based on conditional means that can be used for predicting $Y$ given $X$. Conditional quantiles have been studied in various models (see (Ferraty et al. 2010 [122]) for recent advances and references therein). Prediction can be also performed by means of a conditional mode (see e.g., (Ezzahrioui & Ould Saïd 2008 [115]) and references therein). Conditional distribution with a functional covariate has much potential impact, like in hazard estimation (see e.g., (Quintela del Rio 2008 [265]) and (Ferraty et al. 2010 [122])) or robust regression estimation (see e.g., (Crambes et al. [76])). See (Ferraty et al. 2010 [122]) for a general unifying presentation of nonparametric estimation of various nonlinear operators.

## 17.4.2   Other Types of Regression Problems

This question of predicting $Y$ given $X$ may be asked in other contexts than the specific case of scalar response discussed before in Section 17.4.1, even if the theoretical literature is much more limited. The situation when the response $Y$ is functional and the explanatory one $X$ is multivariate has been mainly studied through applied situations (see for instance (Brumback & Rice 1998 [58]) and Chapter 13 in (Ramsay & Silverman 2005 [267])). The prediction problem when both the response $Y$ and the explanatory variable $X$ are functional has been little studied: the most recent theoretical advances can be found in (Yao et al. 2005 [318]). Another important functional regression setting deals with multilevel functional data. Such statistical situations occur when one observes a functional response in which one is interested in taking into account group-effect, subject-effect within groups, unit-effect within subjects, etc., which encompasses functional multiway analysis of variance (ANOVA) (see (Staicu et al. 2010 [288]) for the most recent developments).

## 17.4.3   Discrimination of Functional Data

Discrimination problems can be viewed as prediction problems when the response $Y$ takes values in a finite set (corresponding to the various groups), such as for instance in the speech recognition example presented in (Delsol et al. 2011 [85]). In a linear context, this question can be directly treated by means of spectral analysis and references in this sense were already discussed before in Section 17.2.2. Chapter 8 in (Ferraty & Vieu 2006 [126])) discusses how functional discrimination can be seen as a special case of nonparametric regression with scalar response, and finally it turns out that all the nonparametric advances discussed before in Section 17.4.1 can be applied to functional discrimination. In addition, a few papers develop specific tools for nonparametric functional discrimination, such as for instance (Ferraty & Vieu 2006 [126]), (Abraham et al. 2006 [1]), (Ferré & Yao 2005 [128])), and (Preda et al. 2007 [261]).

### 17.4.4 Functional Time Series Analysis

When analyzing a time series by means of the functional approach (see example in (Delsol et al. 2011 [85])) one has to deal with a prediction problem with functional covariate $X$ (a continuous path in the past) and a response $Y$ corresponding to some characteristic (functional or not) of the future. The problem cannot be seen as a special case of those discussed in Sections 17.4.1 and 17.4.2, but needs suitable adaptation for taking into account the dependence between the pairs $(X_i, Y_i)_{i=1,...,n}$. General discussions are provided by (Bosq & Blanke 2007 [34]) for the linear approach and in part III of (Ferraty & Vieu 2006 [126]) for nonparametric ones. A very recent and complete bibliographical survey is provided in chapter 5 of (Ferraty & Romain 2011 [124]).

## 17.5 Complementary Bibliography

### 17.5.1 On Testing Problems in FDA

The literature concerning testing problems did not yet receive all the attention that could be expected. This statistical area is certainly one of the main challenges for the statistical community in the future. See for instance (Cuevas et al. 2004 [80]) and (Bugni et al. 2009 [59]) for curve comparison problems, and (Gabrys et al. 2010 [138]) and (Delsol et al. 2011 [86]) and references therein for testing procedures in regression problems. Practical feasibility of such testing procedures is mainly linked with the ability to develop resampling procedures, and a sample of recent advances include (Ferraty et al. 2010 [125]) and (González Manteiga & Martinez Calvo 2011 [146]) in relation to bootstrapping in FDA. See chapter 7 of (Ferraty & Romain 2011 [124]) for a larger discussion on bootstrapping and resampling ideas in FDA.

### 17.5.2 Links with Longitudinal Data Analysis

This discussion has deliberately omitted longitudinal data analysis, which is also another important field of statistics. The reason is because one usual feature of longitudinal data is to be composed of very sparsely discretized curves for which statistical techniques may appreciably differ from those discussed here, even if this is not always true and sometimes FDA methods can be used in longitudinal analysis. The reader may find some elements for comparing longitudinal and functional data approaches in the special issue of the journal *Statistica Sinica* (2004, vol. 14).

## 17.6 Conclusions

This bibliographical survey attests that FDA is a very active field of modern statistics. Many questions remain to be treated and we hope that this discussion has contributed to highlight the main issues. We would like to close this discussion by pointing out some of the main challenges that FDA proposes now to the statistical community.

First, because of the youth of FDA, semiparametric modelling ideas are still underdeveloped compared with parametric or with pure nonparametric approaches. Because of their usefulness in multivariate statistical problems, semiparametric ideas are obviously expected to be of highest interest in infinite-dimensional situations and hence will surely play a prominent role in the future. In the same spirit, the testing problems are quite underdeveloped compared with estimation techniques. Even more than in standard high (but finite) dimensional studies, the structural complexity of functional data makes it difficult to decide whether one specific model is suitable or not. So, the construction of testing procedures adapted to FDA problems is a quite virgin field that should receive great attention in the future. Also it is the construction of specific resampling procedures for dealing with functional datasets that will be a necessary step for practical use of sophisticated methodologies (such as the testing procedures or confidence sets constructions). There are also important fields related with learning theory, receiver operating characteristic (ROC) curves, spatial statistics, ..., whose links with FDA will certainly deserve attention in the future.

## Acknowledgments

The authors wish to thank the participants of the working group STAPH in Toulouse (http://www.math.univ-toulouse.fr/staph/). Special thanks go to Laurent Delsol, Frédéric Ferraty, and Adela Martinez Calvò who have actively collaborated in this survey work. The research of the first author was supported by the national project MTM2008-03010 and the regional project PGIDIT07PXIB207031PR.

# Bibliography

[1] C. Abraham, G. Biau, and B. Cadre. On the kernel rule for function classification. *Annals of the Institute for Statistical Mathematics*, 58(3):619–633, 2006.

[2] C. C. Aggarwal, J. Han, J. Wang, and P. S. Yu. A framework for clustering evolving data streams. In *VLDB'2003: Proceedings of the 29th International Conference on Very Large Data Bases*, pages 81–92. VLDB Endowment, 2003.

[3] A. Ait Saidi, F. Ferraty, R. Kassa, and P. Vieu. Cross-validated estimations in the single functional index model. *Statistics*, 42:475–494, 2008.

[4] J. H. Albert and S. Chib. Bayesian analysis of binary and polychotomous response data. *Journal of the American Statistical Association*, 88(422):669–679, 1993.

[5] M. S. Aldenderfer and R. K. Blashfield. *Cluster Analysis*. Sage Publications, Beverly Hills, California, 1984.

[6] G. Aneiros Perez and P. Vieu. Nonparametric time series prediction: a semi-functional partial linear modeling. *Journal of Multivariate Analysis*, 99:834–857, 2008.

[7] P. Arno, A. G. De Volder, A. Vanlierde, M.-C. Wanet-Defalque, E. Streel, A. Robert, S. Sanabria-Bohórquez, and C. Veraart. Occipital activation by pattern recognition in the early blind using auditory substitution for vision. *NeuroImage*, 13:632–645, 2001.

[8] A. Ashtekar and C. Rovelli. A loop representation for the quantum Maxwell field. *Classical and Quantum Gravity*, 9:1121–1150, 1992.

[9] J. A. D. Aston, J. M. Chiou, and J. E. Evans. Linguistic pitch analysis using functional principal component mixed effect models. *Journal of the Royal Statistical Society Series C*, 59:297–317, 2010.

[10] S. Asur and B. A. Huberman. Predicting the future with social media. In *Proceedings of the 2010 IEEE/WIC/ACM International Conference on Web Intelligence and Intelligent Agent Technology (WI-IAT)*, pages 492–499, 2010.

[11] V. Baladandayuthapani, B. K. Mallick, M. Y. Hong, J. R. Lupton, N. D. Turner, and R. J. Carroll. Bayesian hierarchical spatially correlated functional data analysis with application to colon carcinogenesis. *Biometrics*, 64:64–73, 2008.

[12] A. Bar-Hen, M. Mariadassou, M. A. Poursat, and P. Vandenkoornhuyse. Influence function for robust phylogenetic reconstructions. *Molecular Biology and Evolution*, 25(5):869–873, May 2008.

[13] A. L. Barabasi. *Linked*. Plume, Penguin Group, 2002.

[14] J. Barrientos Marin, F. Ferraty, and P. Vieu. Locally modelled regression and functional data. *Journal of Nonparametric Statistics*, 22(5-6):617–632, 2010.

[15] M. Belkin and P. Niyogi. Semi-supervised learning on Riemannian manifolds. *Machine Learning*, 56:209–239, 2004.

[16] K. Benhenni, F. Ferraty, M. Rachdi, and P. Vieu. Local smoothing regression with functional data. *Computational Statistics*, 22:353–369, 2007.

[17] J.-P. Benzécri. *L'Analyse des Données, Tome I Taxinomie, Tome II Correspondances*. Dunod, 1979. 1st edition, 1973, 2nd edition 1979.

[18] J. P. Benzécri. *Histoire et Préhistoire de l'Analyse des Données*. Dunod, 1982.

[19] J. P. Benzécri. *Correspondance Analysis Handbook*. Marcel Dekker, 1992. Translated by T.K. Gopalan.

[20] J. P. Benzécri. *Linguistique et Lexicologie*. Dunod, 1993.

[21] J. P. Benzécri and F. Benzécri. *Analyse des Correspondances. Exposé Elémentaire*. Dunod, 1980.

[22] J. P. Benzécri and F. Benzécri. *Pratique de l'Analyse des Données, Tome 5: Economie*. Dunod, 1986.

[23] J. P. Benzécri, F. Benzécri, and G. D. Maïti. *Pratique de l'Analyse des Donnés, Tome 4: Médicine, Pharmacologie, Physiologie Clinique*. Statmatic, 1992.

[24] L. Billard and E. Diday. *Symbolic Data Analysis: Conceptual Statistics and Data Mining*. Computational Statistics. Wiley-Interscience, New York, 2006.

[25] L. J. Billera, S. P. Holmes, and K. Vogtmann. Geometry of the space of phylogenetic trees. *Advances in Applied Mathematics*, 27:733–767, 2001.

[26] R. G. Bittar, M. Ptito, J. Faubert, S. O. Dumoulin, and A. Ptito. Activation of the remaining hemisphere following stimulation of the blind hemifield in hemispherectomized subjects. *NeuroImage*, 10:339–346, 1999.

[27] V. D. Blondel, J. L. Guillaume, R. Lambiotte, and E. Lefebvre. Fast unfolding of communities in large networks. *Journal of Statistical Mechanics*, 2008:P10008, October 2008.

[28] A. Blum and T. Mitchell. Combining labeled and unlabeled examples with co-training. In *COLT, Conference on Learning Theory*, pages 92–100, 1998.

[29] H. H. Bock and E. Diday. *Analysis of Symbolic Data: Exploratory Methods for Extracting Statistical Information from Complex Data*. Springer-Verlag, Heidelberg, 2000.

[30] A. Bordes, L. Bottou, and P. Gallinari. SGD-QN: Careful quasi-Newton stochastic gradient descent. *Journal of Machine Learning Research*, 10:1737–1754, July 2009. With erratum, JMLR 11:2229–2240, 2010.

[31] A. Bordes, S. Ertekin, J. Weston, and L. Bottou. Fast kernel classifiers with online and active learning. *Journal of Machine Learning Research*, 6:1579–1619, 2005.

[32] M. Bordewich, A. G. Rodrigo, and C. Semple. Selecting taxa to save or sequence: desirable criteria and a greedy solution. *Systematic Biololgy*, 57(6):825–834, Dec 2008.

[33] M. Börjesson. *Transnationella utbildningsstrategier vid svenska lärosäten och bland svenska studenter i Paris och New York*. PhD thesis, Uppsala universitet, 2005. Ak. avh., Disputationsupplaga, Rapporter från Forskningsgruppen för utbildnings- och kultursociologi 37, SEC/ILU.

[34] D. Bosq and D. Blanke. *Inference and Prediction in Large Dimensions*. Wiley, 2007.

[35] L. Bottou. Online algorithms and stochastic approximations. In David Saad, editor, *Online Learning and Neural Networks*. Cambridge University Press, Cambridge, UK, 1998.

[36] L. Bottou and O. Bousquet. The tradeoffs of large scale learning. In J.C. Platt, D. Koller, Y. Singer, and S. Roweis, editors, *Advances in Neural Information Processing Systems*, volume 20, pages 161–168. NIPS Foundation (http://books.nips.cc), 2008.

[37] L. Bottou and Y. LeCun. On-line learning for very large datasets. *Applied Stochastic Models in Business and Industry*, 21(2):137–151, 2005.

[38] L. Bottou and C. J. Lin. Support Vector Machine solvers. In *Large Scale Kernel Machines*, pages 1–27. 2007.

[39] P. Bourdieu. *Sociologie de l'Algérie*. PUF, 1958.

[40] P. Bourdieu. Champ intellectuel et projet créateur. *Les Temps modernes*, (246):865–906, 1966.

[41] P. Bourdieu. Le marché des biens symboliques. *L'Année sociologique*, 22:49–126, 1971.

[42] P. Bourdieu. *La distinction. Critique sociale du jugement*. Minuit, 1979.

[43] P. Bourdieu. *Homo academicus*. Minuit, 1984.

[44] P. Bourdieu. *La noblesse d'Etat. Grandes écoles et esprit de corps*. Minuit, 1989.

[45] P. Bourdieu. Une révolution conservatrice dans l'édition. *Actes de la recherche en sciences sociales*, 126–127:3–28, 1999.

[46] P. Bourdieu. *Les structures sociales de l'économie*. Seuil, 2000.

[47] P. Bourdieu. Science de la science et réflexivité. *Raisons d'agir*, 2001.

[48] P. Bourdieu and A. Darbel. *L'amour de l'art*. Minuit, 1966.

[49] P. Bourdieu and M. De Saint-Martin. L'anatomie du goût. *Actes de la recherche en sciences sociales*, 2(5):18–43, 1976.

[50] P. Bourdieu and M. De Saint-Martin. Le patronat. *Actes de la recherche en sciences sociales*, 20–21:3–82, 1978.

[51] P. Bourdieu et al. L'économie de la maison. *Actes de la recherche en sciences sociales*, 81–82, 1990.

[52] P. Bourdieu and J.-C. Passeron. *Les héritiers. Les étudiants et la culture*. Minuit, 1964.

[53] P. Bourdieu, A. Sayad, A. Darbel, and C. Seibel. *Travail et travailleurs en Algérie*. Minuit, 1963.

[54] O. Bousquet. *Concentration Inequalities and Empirical Processes Theory Applied to the Analysis of Learning Algorithms*. PhD thesis, Ecole Polytechnique, Palaiseau, France, 2002.

[55] L. Breiman. Bagging predictors. *Machine Learning*, 24:123–140, 1996.

[56] L. Breiman. Random forests. *Machine Learning*, 45:5–32, 2001.

[57] L. Breiman, J. H. Friedman, R. A. Olshen, and C. J. Stone. *Classification and Regression Trees*. Belmont, CA: Wadsworth, 1984.

[58] B. Brumback and J. Rice. Smoothing spline models for the analysis of nested and crossed curves. *Journal of the American Statistical Association*, 93:961–994, 1998.

[59] F. A. Bugni, P. G. Hall, J. L. Horowitz, and G. R. Neumann. Goodness-of-fit tests for functional data. *Econometrics J.*, 12:1–18, 2009.

[60] F. Burba, F. Ferraty, and P. Vieu. k-Nearest neighbor method in functional nonparametric regression. *Journal of Nonparametric Statistics*, 21:453–469, 2009.

[61] M. Burnett and J. Allison. Everybody comes to Rick's, 1940. Screenplay.

[62] K. P. Burnham and D. R. Anderson. *Model Selection and Multimodel Inference: A Practical Information-Theoretical Approach*. Springer-Verlag, 2 edition, 2002.

[63] H. Cardot, A. Mas, and P. Sarda. CLT in functional linear regression models. *Probability Theory and Related Fields*, 138:325–361, 2007.

[64] H. Cardot and P. Sarda. Linear regression models for functional data. In S. Sperlich, W. Haerdle, and G. Aydinli, editors, *The Art of Semiparametrics*, pages 49–66. Physica-Verlag, Heidelberg, 2006.

[65] L. L. Cavalli-Sforza and A. W. F. Edwards. Phylogenetics analysis: Models and estimation procedures. *American Journal of Human Genetics*, 19:233–257, 1967.

[66] G. Celeux, E. Diday, G. Govaert, Y. Lechevallier, and H. Ralambondrainy. *Classification automatique des données*. Dunod, Paris, 1989.

[67] C. C. Chang and C. J. Lin. LIBSVM: a library for Support Vector Machines, 2001. Available at http://www.csie.ntu.edu.tw/~cjlin/libsvm.

[68] G. Chatelet. *Figuring Space: Philosophy, Mathematics, and Physics*. Kluwer, 2000. Translated by R. Shaw and M. Zagha.

[69] J. M. Chiou and H. G. Müller. Diagnostics for functional regression via residual processes. *Computational Statistics and Data Ananysis*, 51(10):4849–4863, 2007.

[70] J. M. Chiou and H. G. Müller. Modeling hazard rates as functional data for the analysis of cohort lifetables and mortality forecasting. *Journal of the American Statistical Association*, 104:572–585, 2009.

[71] A. Ciampi, A. Couturier, and Li. Shaolin. Prediction trees with soft nodes for binary outcomes. *Statistics in Medicine*, 21:1145–1165, 2002.

[72] A. Ciampi, E. Diday, J. Lebbe, E. Perinel, and R. Vignes. Growing a tree classifier with imprecise data. *Pattern Recognition Letters*, 21:787–803, 2000.

[73] A. Clauset, C. Moore, and M. E. J. Newman. Hierarchical structure and the prediction of missing links in networks. *Nature*, 453:98–101, 2008.

[74] A. Clauset, M. E. J. Newman, and C. Moore. Finding community structure in very large networks. *Physical Review E*, 70:066111, 2004.

[75] C. Cortes and V. Vapnik. Support-vector network. *Machine Learning*, 20(3):273–297, 1995.

[76] C. Crambes, L. Delsol, and A. Laksaci. Robust nonparametric estimation for functional data. *Journal of Nonparametric Statistics*, 20:573–598, 2008.

[77] C. Crambes, A. Kneip, and P. Sarda. Smoothing splines estimators for functional linear regression. *Annals of Statisics*, 22(3):35–72, 2009.

[78] B. Csernel, F. Clerot, and G. Hebrail. Streamsamp: Datastream clustering over tilted windows through sampling. In *ECML PKDD 2006 Workshop on Knowledge Discovery from Data Streams*, 2006.

[79] A. J. Cuesta and R. Fraiman. Impartial trimmed k-means for functional data. *Computational Statistics and Data Analysis*, 51:4864–4877, 2007.

[80] A. Cuevas, M. Febrero, and R. Fraiman. An ANOVA test for functional data. *Computational Statistics and Data Analysis*, 47:111–122, 2004.

[81] Darras. *Le partage des bénéfices*. Minuit, 1966.

[82] C. Darwin. *On the Origin of Species by Means of Natural Selection, or the Preservation of Favoured Races in the Struggle for Life*. John Murray, London, 1st edition, 1859.

[83] A. G. De Volder, H. Toyama, Y. Kimura, M. Kiyosawa, H. Nakano, A. Vanlierde, M.-C. Wanet-Defalque, M. Mishina, K. Oda, K. Ishiwata, and M. Senda. Auditory triggered mental imagery of shape involves visual association areas in early blind humans. *NeuroImage*, 14:129–139, 2001.

[84] P. Delicado. Functional k-sample problem when data are density functions. *Computational Statistics*, 22(3):391–410, 2007.

[85] L. Delsol, F. Ferraty, and A. Martinez Calvo. Functional data analysis: an interdisciplinary statistical topic. In *Statistical Learning and Data Science*. Chapaman & Hall, 2011. This volume.

[86] L. Delsol, F. Ferraty, and P. Vieu. Structural tests in regression on functional variables. *Journal of Multivariate Analysis*, 102:422–447, 2011.

[87] A. P. Dempster, N. M. Laird, and D. B. Rubin. Maximum likelihood estimation from incomplete data via the EM algorithm (with discussion). *Journal of the Royal Statistical Society, Series B*, 39(1):1–38, 1977.

[88] J. E. Dennis and R. B. Schnabel. *Numerical Methods For Unconstrained Optimization and Nonlinear Equations*. Prentice-Hall, Inc., Englewood Cliffs, New Jersey, 1983.

[89] F. Denord. *Genèse et institutionnalisation du néo-libéralisme en France (années 1930–1950).* PhD thesis, EHESS, 2003.

[90] A. Desrosières. Une rencontre improbable et deux héritages. In P. Encrevé and R.-M. Lagrave, editors, *Travailler avec Bourdieu*, pages 209–218. Flammarion, 2003.

[91] A. Desrosières. Analyse des données et sciences humaines: comment cartographier le monde social ? *Journal Electronique d'Histoire des Probabilités et de la Statistique*, pages 11–19, 2008.

[92] C. Di, C. M. Crainiceanu, B.S. Caffo, and N.M. Punjabi. Multilevel functional principal component analysis. *Annals of Applied Statistics*, 3:458–488, 2009.

[93] E. Diday. La méthode des nuées dynamiques. *Revue de Statistique Appliquée*, XIX(2):19–34, 1971.

[94] E. Diday. Introduction à l'approche symbolique en analyse des données. In *Actes des Premières Journées Symbolique-Numérique*, Université Paris IX Dauphine, Paris, France, 1987.

[95] E. Diday. The symbolic approach in clustering and related methods of data analysis. In *The Basic Choices, IFCS-87*, pages 673–684, 1988.

[96] E. Diday. Introduction à l'analyse des données symbolique. *Rapport de recherche INRIA*, XIX(1074), 1989.

[97] E. Diday. Categorization in symbolic data analysis. In H. Cohen and C. Lefebvre, editors, *Handbook of Categorization in Cognitive Science*, pages 845–867. Elsevier, 2005.

[98] E. Diday and R. Emilion. Treillis de Galois maximaux et capacités de Choquet. *Académie des Sciences Paris Comptes Rendus Série Sciences Mathématiques*, 325:261–266, 1997.

[99] E. Diday and R. Emilion. Maximal and stochastic galois lattices. *Journal of Discrete Applied Mathematics*, 127:271–284, 2003.

[100] E. Diday and M. Noirhomme-Fraiture. *Symbolic Data Analysis and the SODAS Software.* Wiley-Interscience, New York, NY, USA, 2008.

[101] P. D'Urso and M. Vichi. Dissimilarities between trajectories of a three-way longitudinal data set. In *Advances in Data Science and Classification*, pages 585–592. Springer-Verlag, Heidelberg, 1998.

[102] J. Duval. *Critique de la raison journalistique. Les transformations de la presse économique en France.* Seuil, 2004.

[103] A. W. F. Edwards and L. L. Cavalli-Sforza. The reconstruction of evolution. *Annals of Human Genetics*, 27:105–106, 1963.

[104] A. W. F. Edwards and L. L. Cavalli-Sforza. Phenetic and phylogenetic classification. chapter Reconstruction of evolutionary trees, pages 67–76. Systematics Association Publ. No. 6, London, 1964.

[105] J. P. Egan. *Signal Detection Theory and ROC Analysis.* Academic Press, 1975.

[106] J. A. Eisen. Phylogenomics: improving functional predictions for uncharacterized genes by evolutionary analysis. *Genome Research*, 8(3):163–167, Mar 1998.

[107] J. A. Eisen and M. Wu. Phylogenetic analysis and gene functional predictions: phylogenomics in action. *Theoretical Population Biology*, 61(4):481–487, Jun 2002.

[108] O. Elemento. Apport de l'analyse en composantes principales pour l'initialisation et la validation de cartes topologiques de Kohonen. In *SFC'99, Journées de la Société Francophone de la Classification*, Nancy, France, 1999.

[109] J. Eliashberg, A. Elberse, and M.A.A.M. Leenders. The motion picture industry: critical issues in practice, current research, and new research directions. *Marketing Science*, 25:638–661, 2006.

[110] J. Eliashberg, S.K. Hui, and Z.J. Zhang. From storyline to box office: a new approach for green-lighting movie scripts. *Management Science*, 53:881–893, 2007.

[111] P. Erdős and A. Rényi. On random graphs. *Publicationes Mathematicae*, 6:290–297, 1959.

[112] M. Escabias, A. M. Aguilera, and M. Valderrama. Principal component estimation of functional logistic regression: discussion of two different approaches. *Journal of Nonparametric Statistics*, 16:365–384, 2004.

[113] M. Escabias, A. M. Aguilera, and M. Valderrama. Modeling environmental data by functional principal component logistic regression. *Environmetrics*, 16(1):95–107, 2005.

[114] B. Everitt, S. Landau, and M. Leese. *Cluster Analysis*. Oxford University Press, 2001.

[115] M. Ezzahrioui and E. Ould Saïd. Asymptotic normality of nonparametric estimator of the conditional mode. *Journal of Nonparametric Statistics*, 20:3–18, 2008.

[116] R. E. Fan, P. H. Chen, and C. J. Lin. Working set selection using second order information for training SVM. *Journal of Machine Learning Research*, 6:243–264, 2005.

[117] J. Farquhar, D. R. Hardoon, H. Meng, J. Shawe-Taylor, and S. Szedmak. Two view learning: SVM-2K, theory and practice. In *NIPS, Neural Information Processing Systems*, pages 355–362, 2005.

[118] J. Felsenstein. Maximum likelihood and minimum-steps methods for estimating evolutionary trees from data on discrete characters. *Systematic Zoology*, 22:240–249, 1973.

[119] B. M. Fernández de Castro and W. González Manteiga. Boosting for real and functional samples: an application to an environmental problem. *Stochastic Environmental Research and Risk Assessment*, 22(1):27–37, 2008.

[120] F. Ferraty, W. González Manteiga, A. Martinez Calvo, and P. Vieu. Presmoothing in functional linear regression. *Statistica Sinica*, 2011. (to appear).

[121] F. Ferraty, P. G. Hall, and Vieu P. Most predictive design points in functional predictors. *Biometrika*, 97:807–824, 2010.

[122] F. Ferraty, A. Laksaci, A. Tadj, and P. Vieu. Rate of uniform consistency for non-parametric estimates with functional variables. *Journal of Statistical Planning and Inference*, 140:335–352, 2010.

[123] F. Ferraty, A. Rabhi, and P. Vieu. Conditional quantiles for dependent functional data with application to the climatic El Niño phenomenon. *Sankhyā*, 67(2):378–398, 2005.

[124] F. Ferraty and Y. Romain, editors. *Oxford Handbook of Functional Data Analysis*. Oxford Handbooks. Oxford University Press, 2011.

[125] F. Ferraty, I. Van Keilegom, and P. Vieu. On the validity of the bootstrap in functional regression. *Scandinavian Journal of Statistics*, 37:286–306, 2010.

[126] F. Ferraty and P. Vieu. *Nonparametric Functional Data Analysis: Theory and Practice*. Springer, New York, 2006.

[127] F. Ferraty and P. Vieu. Additive prediction and boosting for functional data. *Computational Statistics and Data Analysis*, 53(10):1400–1413, 2009.

[128] L. Ferré and A. F. Yao. Smoothed functional inverse regression. *Statistica Sinica*, 15(3):665–683, 2005.

[129] W. M. Fitch and E. Margoliash. Construction of phylogenetic trees. *Science*, 155(760):279–284, Jan 1967.

[130] F. Fogelman Soulié. Modeling with networked data. In *KDD 2010, 16th ACM SIGKDD Conference on Knowledge Discovery and Data Mining*, 2010. Invited talk in Data Mining Case Studies Sessions, http://videolectures.net/kdd2010_fogelman_soulie_mnd.

[131] F. Fogelman Soulié and E. Marcadé. Industrial mining of massive data sets. In F. Fogelman-Soulié, D. Perrotta, J. Pikorski, and R. Steinberger, editors, *Mining Massive Data Sets for Security Advances in Data Mining, Search, Social Networks and Text Mining and Their Applications to Security*, NATO ASI Series, pages 44–61. IOS Press, 2008.

[132] F. Fogelman Soulié, T. Porez, D. Ristic, and S. Sean. Utilisation des réseaux sociaux pour la sécurité : un exemple pour la détection et l'investigation de la fraude à la carte bancaire sur Internet. In *Proceedings WISG 2011*. ANR, 2011.

[133] S. Fortunato. Community detection in graphs. *Physics Reports*, 486(3–5):75–174, 2010.

[134] S. Fortunato and M. Barthelemy. Resolution limit in community detection. *Proceedings of the National Academy of Sciences*, 104(1):36–41, 2007.

[135] J. L. Foulley and F. Jaffrézic. Modelling and estimating heterogeneous variances in threshold models for ordinal discrete data via winbugs/openbugs. *Computer Methods and Programs in Biomedicine*, 97(1):19–27, 2010.

[136] J. H. Friedman. On bias, variance, 0/1-loss, and the curse of dimensionality. *Data Mining and Knowledge Discovery*, 1:55–77, 1997.

[137] P. Gabrielson. Worldwide trends in lung cancer pathology. *Respirology, 11*, pages 533–538, 2006.

[138] R. Gabrys, L. Horváth, and P. Kokoszka. Tests for error correlation in the functional linear model. *Journal of the American Statistical Association*, 105:1113–1125, 2010.

[139] A. Gammerman, I. Nouretdinov, B. Burford, A. Y. Chervonenkis, V. Vovk, and Z. Luo. Clinical mass spectrometry proteomic diagnosis by conformal predictors. *Statistical Applications in Genetics and Molecular Biology*, 7(2):art. 13, 2008.

[140] A. Gammerman and V. Vovk. Hedging predictions in machine learning (with discussion). *Computer Journal*, 50(2):151–177, 2007.

[141] M.-F. Garcia-Parpet. Des outsiders dans l'économie de marché. Pierre Bourdieu et les travaux sur l'Algérie. *Awal, numéro spécial "L'autre Bourdieu,"* 27–28:139–150, 2003.

[142] M. Gettler-Summa, L. Schwartz, J. M. Steyaert, F. Vautrain, M. Barrault, and N. Hafner. Multiple time series: new approaches and new tools in data mining applications to cancer epidemiology. *Revue Modulad*, 34:37–46, 2006.

[143] M. Girvan and M. E. J. Newman. Community structure in social and biological networks. *Proceedings of the National Academy of Sciences*, 99(12):7821–7826, 2002.

[144] M. Gladwell. The formula: what if you built a machine to predict hit movies? *The New Yorker*, 16 Oct. 2006. www.newyorker.com/archive/2006/10/16/061016fa_fact6.

[145] M. Gönen. Analyzing receiver operating characteristic curves with SAS. Cary, 2007. SAS Institute Inc., North Carolina.

[146] W. González Manteiga and A. Martinez Calvo. Bootstrap in functional linear regression. *Journal of Statistical Planning and Inference*, 141(1):453–461, 2011.

[147] W. González Manteiga and P. Vieu. Methodological richness of functional data. In *Statistical Learning and Data Science*. Chapman & Hall, 2011. This volume.

[148] A. D Gordon. *Classification*. Chapman and Hall/CRC, Boca Raton, Florida, 1999.

[149] M. Guillaumin, T. Mensink, J. Verbeek, and C. Schmid. Tagprop: Discriminative metric learning in nearest neighbor models for image auto-annotation. In *Proceedings of ICCV, International Conference on Computer Vision*, 2009.

[150] S. Guindon and O. Gascuel. A simple, fast, and accurate algorithm to estimate large phylogenies by maximum likelihood. *Systematic Biology*, 52(5):696–704, Oct 2003.

[151] P. G. Hall, P. Poskitt, and D. Presnell. A functional data-analytic approach to signal discrimination. *Technometrics*, 43:1–9, 2001.

[152] P. G. Hall and C. Vial. Assessing the finite dimensionality of functional data. *Journal of the Royal Statistical Society Series B*, 68(4):689–705, 2006.

[153] D. J. Hand. Assessing classification rules. *Journal of Applied Statistics*, 21:3–16, 1994.

[154] D. J. Hand. *Construction and Assessment of Classification Rules*. Wiley, 1997.

[155] D. J. Hand. Breast cancer diagnosis from proteomic mass spectrometry data: a comparative evaluation. *Statistical Applications in Genetics and Molecular Biology*, 7(2):art. 15, 2008.

[156] D. J. Hand. Measuring classifier performance: a coherent alternative to the area under the ROC curve. *Machine Learning*, 77:103–123, 2009.

[157] D. J. Hand. Evaluating diagnostic tests: the area under the ROC curve and the balance of errors. *Statistics in Medicine*, 29:1502–1510, 2010.

[158] D. J. Hand. A note on interpreting the area under the ROC curve. *Technical Report*, 2010.

[159] D.J. Hand and F. Zhou. Evaluating models for classifying customers in retail banking collections. *Journal of the Operational Research Society*, 61:1540–1547, 2010.

[160] W. Härdle, Y. Mori, and P. Vieu, editors. *Statistical Methods for Biostatistics and Related Fields*. Contributions to Statistics. Springer-Verlag, Heidelberg, 2007.

[161] T. Hastie, A. Buja, and R. Tibshirani. Penalized discriminant analysis. *Annals of Statistics*, 13:435–475, 1995.

[162] T. Hastie, R. Tibshirani, and J. Friedman. *The Elements of Statistical Learning: Data Mining, Inference, and Prediction*. Springer, 2001.

[163] G. He, H. G. Müller, and J. L. Wang. Methods of canonical analysis for functional data. *Journal of Statistical Planning and Inference*, 122(1-2):141–159, 2004.

[164] N. E. Heckman and R. H. Zamar. Comparing the shapes of regression functions. *Biometrika*, 87(1):135–144, 2000.

[165] J. Hjellbrekke, B. Le Roux, O. Korsnes, F. Lebaron, L. Rosenlund, and H. Rouanet. The Norwegian fields of power anno 2000. *European Societies*, 9(2):245–273, 2007.

[166] D. Hlubinka and L. Prchal. Changes in atmospheric radiation from the statistical point of view. *Computational Statistics and Data Analysis*, 51(10):4926–4941, 2007.

[167] B. Hotu, E. Diday, and M. Summa. Generating rules for expert system from observations. *Pattern Recognition Letters*, 7:271–276, 1988.

[168] J. F. Hovden. *Profane and Sacred: A Study of the Norwegian Journalistic Field*. PhD thesis, University of Bergen, Faculty of the Social Sciences, 2008.

[169] T. Hsing and H. Ren. An RKHS formulation of the inverse regression dimension reduction problem. *Annals of Statistics*, 37:726–755, 2009.

[170] B. A. Huberman. *The Laws of the Web: Patterns in the Ecology of Information*. MIT Press, 2001.

[171] L. Hubert and P. Arabie. Comparing partitions. *Journal of Classification*, 2:193–218, 1985.

[172] R. J. Hyndman and S. Ullah. Robust forecasting of mortality and fertility rates: a functional data approach. *Computational Statistics and Data Analysis*, 51(10):4942–4957, 2007.

[173] ISO/TC69. Méthodes statistiques utilisées dans les essais d'aptitude par comparaisons interlaboratoires. ISO 13528:2005. International Organization for Standardization (ISO), Geneva, Switzerland, 2005.

[174] R. Izmailov, V. Vapnik, and A. Vashist, 2009. Private communication.

[175] A. Jain, M. Murty, and P. Flynn. Data clustering: a review. *ACM Computing Survey*, 31(3):264–323, 1999.

[176] A. Jamain. *Meta Analysis of Classification Methods*. PhD thesis, Department of Mathematics, Imperial College, London, 2004.

[177] A. Jamain and D.J. Hand. Mining supervised classification performance studies: a meta-analytic investigation. *Journal of Classification*, 25:87–112, 2008.

[178] G. M. James and T. Hastie. Functional linear discriminant analysis for irregularly sampled curves. *Journal of the Royal Statistical Society Series B*, 63:533–550, 2001.

[179] G. M. James and C. A. Sugar. Clustering for sparsely sampled functional data. *Journal of the American Statistical Association*, 98:397–408, 2003.

[180] B. J. Jansen, M. Zhang, K. Sobel, and A. Chowdury. Twitter power: tweets as electronic word of mouth. *Journal of the American Society for Information Science and Technology*, 60(11):2169–2188, 2009.

[181] C. R. Jiang, J. A. D. Aston, and J. L. Wang. Smoothing dynamic position emission tomography times courses using functional principal components. *Neuroimage*, 47:184–193, 2009.

[182] T. Joachims. Training linear SVMs in linear time. In *Proceedings of the 12th ACM SIGKDD International Conference*, New York, 2006.

[183] Y. Kitazoe, H. Kishino, P. J. Waddell, N. Nakajima, T. Okabayashi, T. Watabe, and Y. Okuhara. Robust time estimation reconciles views of the antiquity of placental mammals. *PLoS ONE*, 2(4):e384, 2007.

[184] J. Kleinberg. Authoritative sources in a hyperlinked environment. *Journal of the ACM*, 46:604–632, 1999.

[185] J. Kleinberg and S. Lawrence. The structure of the web. *Science*, 294(5548):1849–1850, 2001.

[186] J. M. Kleinberg, R. Kumar, P. Raghavan, S. Rajagopalan, and A. S. Tomkins. The web as a graph: measurements, models, and methods. In *Proceedings of COCOON '99, 5th Annual International Conference on Computing and Combinatorics*, pages 1–17. Springer, 1999.

[187] A. Kneip, X. Li, K.B. MacGibbon, and J. O. Ramsay. Curve registration by local regression. *Canadian Journal of Statistics*, 28(1):19–29, 2000.

[188] Y. Kodratoff. *Leçons d'Apprentissage Symbolique Automatique*. Cepadues-Editions, Toulouse, France, 1986.

[189] T. Kohonen. *Self-Organizing Maps*, volume 30 of *Springer Series in Information Sciences*. Springer, third edition, 1995. Most recent edition published in 2001.

[190] M. Kriváanek and J. Morávek. NP-hard problems in hierarchical-tree clustering. *Acta Informatica, 23*, 23:311–323, 1986.

[191] W. J. Krzanowski and D. J. Hand. *ROC Curves for Continuous Data*. Chapman & Hall, 2009.

[192] J. Kumpula. *Community Structures in Complex Networks: Detection and Modeling*. PhD thesis, Helsinki University of Technology, Faculty of Information and Natural Sciences, Department of Biomedical Engineering and Computational Science Publications, Report A06, 2008.

[193] A. Kupresanin, H. Shin, D. King, and R.L. Eubank. An RKHS framework for functional data analysis. *Journal of Statistical Planning and Inference*, 140:3627–3637, 2010.

[194] J. D. Lafferty, A. McCallum, and F. C. N. Pereira. Conditional random fields: Probabilistic models for segmenting and labeling sequence data. In C. E. Brodley and A. P. Danyluk, editors, *Proceedings of the Eighteenth International Conference on Machine Learning (ICML 2001)*, pages 282–289, Williams College, Williamstown, 2001. Morgan Kaufmann.

[195] E. Law and L. Von Ahn. Input-agreement: a new mechanism for data collection using human computation games. In *CHI, ACM Conference on Human Factors in Computer Systems*, pages 1197–1206, 2009.

[196] B. Le Roux and H. Rouanet. *Geometric Data Analysis: From Correspondence Analysis to Structured Data Analysis*. Kluwer, 2004.

[197] B. Le Roux and H. Rouanet. *Multiple Correspondence Analysis*. Sage Publications, 2010.

[198] F. Lebaron. Economists and the economic order: the field of economists and the field of power in France. *European Societies*, 3(1):91–110, 2001.

[199] F. Lebaron. Economic models against economism. *Theory and Society*, 32(5-6):551–565, 2003.

[200] L. Lebart, A. Salem, and L. Berry. *Exploring Textual Data*. Kluwer, 1998. (L. Lebart and A. Salem, Analyse Statistique des Données Textuelles, Dunod, 1988).

[201] S. Y. Lee. *Structural Equation Modelling: A Bayesian Approach*. Wiley, 2007.

[202] W. S. Lee, P. L. Bartlett, and R. C. Williamson. The importance of convexity in learning with squared loss. *IEEE Transactions on Information Theory*, 44(5):1974–1980, 1998.

[203] J. Leskovec, L. A. Adamic, and B. A. Huberman. The dynamics of viral marketing. *ACM Transactions on the Web (TWEB)*, 1(1):art. 5, 2007.

[204] F. Levi, F. Lucchini, E. Negri, W. Zatonski, P. Boyle, and C. LaVecchia. Trends in cancer mortality in the European Union and accession countries, 1980–2000. *Annals of Oncology*, 15:1425–1431, 2004.

[205] D. D. Lewis, Y. Yang, T. G. Rose, and F. Li. RCV1: A new benchmark collection for text categorization research. *Journal of Machine Learning Research*, 5:361–397, 2004.

[206] S. Li. *Phylogenetic Tree Construction Using Markov Chain Monte Carlo*. PhD thesis, Ohio State University, Ohio, 1996.

[207] Y. Li, N. Wang, and R. Carrol. Generalized functional linear models with interactions. *Journal of the American Statistical Association*, 490:621–633, 2010.

[208] C. J. Lin, R. C. Weng, and S. S. Keerthi. Trust region Newton methods for large-scale logistic regression. In Z. Ghahramani, editor, *Twenty-Fourth International Conference on Machine Learning (ICML 2007)*, pages 561–568. ACM, 2007.

[209] Y. R. Lin, Y. Chi, S. Zhu, H. Sundaram, and B. L. Tseng. Facetnet: a framework for analyzing communities and their evolutions in dynamic networks. In *Proceeding of the 17th International Conference on World Wide Web WWW'08*, pages 685–694, 2008.

[210] B. Liu and H. G. Müller. Estimating derivatives for samples of sparsely observed functions, with application to online auction dynamics. *Journal of the American Statistical Association*, 104:704–717, 2009.

[211] C. Liu, D. B. Rubin, and Y. N. Wu. Parameter expansion to accelerate EM: the PX-EM algorithm. *Biometrika*, 85:755–770, 1998.

[212] J. S. Liu and Y. N. Wu. Parameter expansion for data augmentation. *Journal of the American Statistical Association*, 94(448):1264–1274, 1999.

[213] X. Liu and M. C. K. Yang. Simultaneous curve registration and clustering for functional data. *Computational Statistics and Data Analysis*, 53(4):1361–1376, 2009.

[214] N. Locantore, S. J. Marron, D. G. Simpson, N. Tripoli, J. T. Zhang, and K. L. Cohen. Robust component analysis for functional data (with discussion). *TEST*, 8:1–73, 1999.

[215] S. López-Pintado and J. Romo. Depth-based inference for functional data. *Computational Statistics and Data Analysis*, 51(10):4957–4968, 2007.

[216] S. López-Pintado and J. Romo. On the concept of depth for functional data. *Journal of the American Statistical Association*, 104:718–734, 2009.

[217] M. M. De Choudhury, W. A. Mason, J. M. Hofman, and D. J. Watts. Inferring relevant social networks from interpersonal communication. In *WWW 2010: Proceedings of the 19th International Conference on the World Wide Web*, pages 301–310. ACM, 2010.

[218] J. MacQueen. Some methods for classification and analysis of multivariate observations. In L. M. LeCam and J. Neyman, editors, *Proceedings of the Fifth Berkeley Symposium on Mathematics, Statistics, and Probabilities*, volume 1, pages 281–297, Berkeley and Los Angeles, (Calif), 1967. University of California Press.

[219] S. A. Macskassy and F. Provost. Classification in networked data: a toolkit and a univariate case study. *Journal of Machine Learning Research*, 8(5/1):935–983, 2007.

[220] K. Madjid and M. Norwati. Data stream clustering: challenges and issues. In *The 2010 IAENG International Conference on Data Mining and Applications*, Hong Kong, March 2010.

[221] A. R. Mahdiraji. Clustering data stream: a survey of algorithms. *International Journal of Knowledge-Based Intelligent Engineering Systems*, 13(2):39–44, 2009.

[222] M. Mariadassou, A. Bar-Hen, and H. Kishino. Taxon influence index: assessing taxon-induced incongruities in phylogenetic inference. *Systematic Biology*, 2011. In press.

[223] P. Massart. Some applications of concentration inequalities to statistics. *Annales de la Faculté des Sciences de Toulouse*, series 6, 9(2):245–303, 2000.

[224] B. Mau. *Bayesian Phylogenetic Inference Via Markov Chain Monte Carlo*. PhD thesis, University of Wisconsin, 1996.

[225] R. McKee. *Story: Substance, Structure, Style, and the Principles of Screenwriting*. Methuen, 1999.

[226] T. Melluish, C. Saunders, I. Nouretdinov, and V. Vovk. Comparing the Bayes and typicalness frameworks. In *Lecture Notes in Computer Science*, volume 2167, pages 360–371. Springer, Heidelberg, 2001. Full version published as a CLRC technical report TR-01-05; see `http://www.clrc.rhul.ac.uk`.

[227] C. Meza, F. Jaffrézic, and J. L. Foulley. Estimation in the probit normal model for binary outcomes using the SAEM algorithm. *Computational Statistics and Data Analysis*, 53(4):1350–1360, 2009.

[228] R. S. Michalski, J. G. Carbonel, and T. M. Mitchell. *Machine Learning: An Artificial Intelligence Approach*. Tioga Publishing Company, Palo Alto, USA, 1983.

[229] R. S. Michalski, J. G. Carbonel, and T. M. Mitchell. *Machine Learning: An Artificial Intelligence Approach*, volume II. Morgan Kaufmann, Los Altos, USA, 1986.

[230] R. S. Michalski, E. Diday, and R. E. Stepp. A recent advance in data analysis: Clustering objects into classes characterized by conjunctive concepts. In *Progress in Pattern Recognition*, volume I, pages 33–56, North Holland, Amsterdam, 1981.

[231] S. Milgram. The small world problem. *Psychology Today*, 1:61, 1967.

[232] H. G. Müller. Functional modeling of longitudinal data. In Donald Hedeker and Robert D. Gibbons, editors, *Longitudinal Data Analysis*. Wiley, New York, 2009.

[233] N. Murata. A statistical study of on-line learning. In David Saad, editor, *Online Learning and Neural Networks*. Cambridge University Press, Cambridge, UK, 1998.

[234] F. Murtagh. *Multidimensional Clustering Algorithms*. Physica-Verlag, 1985.

[235] F. Murtagh. On ultrametricity, data coding, and computation. *Journal of Classification*, 21:167–184, 2004.

[236] F. Murtagh. *Correspondence Analysis and Data Coding with R and Java*. Chapman & Hall/CRC, 2005.

[237] F. Murtagh. The correspondence analysis platform for uncovering deep structure in data and information (Sixth Boole Lecture). *Computer Journal*, 53:304–315, 2010.

[238] F. Murtagh, A. Ganz, and S. McKie. The structure of narrative: the case of film scripts. *Pattern Recognition*, 42:302–312, 2009. (Discussed in: Z. Merali, Here's looking at you, kid. Software promises to identify blockbuster scripts, *Nature*, 453, 708, 4 June 2008.).

[239] O. Nasraoui, M. Soliman, E. Saka, A. Badia, and R. Germain. A web usage mining framework for mining evolving user profiles in dynamic web sites. *IEEE Transactions on Knowledge and Data Engineering*, 20(2):202–215, 2008.

[240] O. Nasraoui, C. C. Uribe, and C. R. Coronel. Tecno-streams: tracking evolving clusters in noisy data streams with a scalable immune system learning model. In *ICDM 2003: Proceedings of the 3rd IEEE International Conference on Data Mining*, pages 235–242. IEEE Computer Society, Los Alamitos, 2003.

[241] D. Nérini and B. Ghattas. Classifying densities using functional regression trees: application in oceanology. *Computational Statistics and Data Analysis*, 51(10):4984–4993, 2007.

[242] M. E. J. Newman and M. Girvan. Finding and evaluating community structure in networks. *Physical Review E*, 69(2):026113, 2004.

[243] M. A. Nielsen and I. L. Chuang. *Quantum Computation and Quantum Information.* Cambridge University Press, 2000.

[244] J. Nocedal and S. Wright. *Numerical Optimization.* Springer-Verlag, 2nd edition, 2006.

[245] I. Nouretdinov, S. G. Costafreda, A. Gammerman, A. Y. Chervonenkis, V. Vovk, V. Vapnik, and C.H.Y. Fu. Machine learning classification with confidence: Application of transductive conformal predictors to MRI-based diagnostic and prognostic markers in depression. In *Proceedings of Neuroimage*, 2010.

[246] I. Nouretdinov, G. Li, A. Gammerman, and Z. Luo. Application of conformal predictors for tea classification based on electronic nose. In *Proceedings of 6th IFIP International Conference on Artificial Intelligence Applications and Innovations, AIAI*, 2010.

[247] I. Nouretdinov, T. Melluish, and V. Vovk. Ridge regression confidence machine. In *Proceedings of the 18th International Conference on Machine Learning*, pages 385–392, San Mateo, CA, 2001. Morgan Kaufmann.

[248] L. O'Callaghan, N. Mishra, A. Meyerson, S. Guha, and R. Motwani. Streaming-data algorithms for high-quality clustering. In *Proceedings of IEEE International Conference on Data Engineering*, pages 685–694, 2001.

[249] G. Palla, I. Derényi, I. Farkas, and T. Vicsek. Uncovering the overlapping community structure of complex networks in nature and society. *Nature*, 435(7043):814–818, 2005.

[250] S. Pandit, D. H. Chau, S. Wang, and C. Faloutsos. Netprobe: A fast and scalable system for fraud detection in online auction networks. In *Proceedings of the 16th International Conference on World Wide Web*, pages 201–210. ACM, 2007.

[251] J. Park, T. Gasser, and V. Rousson. Structural components in functional data. *Computational Statistics and Data Analysis*, 53(9):3452–3465, 2009.

[252] B. Parker and J. Wen. Predicting microRNA targets in time-series microarray experiments via functional data analysis. *BMC Bioinformatics*, 10(Suppl 1:S32), 2009.

[253] F. Pazos and A. Valencia. Similarity of phylogenetic trees as indicator of protein-protein interaction. *Protein Engineering*, 14(9):609–614, Sep 2001.

[254] J. Peng and H. G. Müller. Distance-based clustering of sparsely observed stochastic processes, with applications to online auctions. *Annals of Applied Statistics*, 2:1056–1077, 2008.

[255] M. S. Pepe. *The Statistical Evaluation of Medical Tests for Classification and Prediction.* Oxford University Press, 2000.

[256] J. Platt. Fast training of Support Vector Machines using Sequential Minimal Optimization. In *Advances in Kernel Methods – Support Vector Learning*, pages 185–208. MIT Press, 1999.

[257] B. T. Polyak and A. B. Juditsky. Acceleration of stochastic approximation by averaging. *SIAM Journal of Control Optimization*, 30(4):838–855, 1992.

[258] A. M. Polyakov. *Gauge Fields and Strings.* Harwood, 1987.

[259] P. Pons and M. Latapy. Computing communities in large networks using random walks. *Journal of Graph Algorithms and Applications*, 10(2):191–218, 2006.

[260] M. J. D Powell. Variable metric methods for constrained optimization. In *Mathematical Programming: The State of the Art*, pages 1425–1431. INRIA, 1983.

[261] C. Preda, G. Saporta, and C. Leveder. PLS classification of functional data. *Computational Statistics*, 22:223–235, 2007.

[262] M. Ptito, P. Johannsen, J. Faubert, and A. Gjedde. Activation of human extrageniculostriate pathways after damage to area V1f. *NeuroImage*, 9:97–107, 1999.

[263] J. R. Quinlan. *Probabilistic decision trees*. Morgan-Kaufmann, 1990.

[264] J. R. Quinlan. *C4.5: Programs for Machine Learning*. Morgan Kaufmann, 1993.

[265] A. Quintela del Rio. Hazard function given a functional variable: nonparametric estimation. *Journal of Nonparametric Statistics*, 20:413–430, 2008.

[266] J. O. Ramsay and B. Silverman. *Applied Functional Data Analysis: Methods and Case Studies*. Springer-Verlag, New York, 2002.

[267] J. O. Ramsay and B. Silverman. *Functional Data Analysis*. Springer-Verlag, New York, 2nd edition, 2005.

[268] B. Rannala and Z. Yang. Probability distribution of molecular evolutionary trees: a new method of phylogenetic inference. *Journal of Molecular Evolution*, 43(3):304–311, Sep 1996.

[269] S. Ressler. Social network analysis as an approach to combat terrorism: past, present, and future research. *Homeland Security Affairs*, II(2), July 2006.

[270] J. Rice and B. Silverman. Estimating the mean and covariance structure nonparametrically when the data are curves. *Journal of the Royal Statistical Society Scries B*, 53(1):233–243, 1991.

[271] H. Robbins and D. Siegmund. A convergence theorem for non negative almost super-martingales and some applications. In Jagdish S. Rustagi, editor, *Optimizing Methods in Statistics*. Academic Press, 1971.

[272] K. Robson and C. Sanders. *Quantifying Theory: Bourdieu*. Springer, 2008.

[273] F. Rosenblatt. The perceptron: A perceiving and recognizing automaton. Technical Report 85-460-1, Project PARA, Cornell Aeronautical Lab, 1957.

[274] L. Rosenlund. *Exploring the City with Bourdieu: Applying Pierre Bourdieu's Theories and Methods to Study the Community*. VDM Verlag, 2009. Foreword by L.Wacquant.

[275] N. Rossi, X. Wang, and J. O. Ramsay. Nonparametric item response function estimates with the EM algorithm. *Journal of the Behavioral and Educational Sciences*, 27:291–317, 2002.

[276] H. Rouanet. The geometric analysis of individuals by variables tables. In M. Greenacre and J. Blasius, editors, *Multiple Correspondence Analysis and Related Methods*. Chapman & Hall, 2006.

[277] H. Rouanet, W. Ackermann, and B. Le Roux. The geometric analysis of question-naires: the lesson of Bourdieu's La distinction. *Bulletin de Méthodologie Sociologique*, (65):5–15, 2000.

[278] D. E. Rumelhart, G. E. Hinton, and R. J. Williams. Learning internal representations by error propagation. In *Parallel Distributed Processing: Explorations in the Microstructure of Cognition*, volume I, pages 318–362. Bradford Books, Cambridge, MA, 1986.

[279] P. F. Russell and T. R. Rao. On habitat and association of species of anophelinae larvae in south-eastern Madras. *Journal of Malaria*, III:153–178, 1940.

[280] S. S. Johnston. *Quantum Fields on Causal Sets*. PhD thesis, Imperial College London, 2010. http://arxiv.org/pdf/1010.5514v1.

[281] G. Sapiro. *La guerre des écrivains (1940–1953)*. Fayard, 1999.

[282] M. Savage, A. Warde, B. Le Roux, and H. Rouanet. Class and cultural division in the UK. *Sociology*, 42(6):1049–1071, 2008.

[283] C. Seibel. Travailler avec Bourdieu. *Awal, numéro spécial "L'autre Bourdieu,"* 27–28:91–97, 2005.

[284] S. Shalev-Shwartz and N. Srebro. SVM optimization: inverse dependence on training set size. In *Proceedings of the 25th International Machine Learning Conference (ICML 2008)*, pages 928–935. ACM, 2008.

[285] A. Skrondal and S. Rabe-Hesketh. *Generalized Latent Variable Modeling*. Chapman & Hall/CRC, 2004.

[286] M. Spiliopoulou, I. Ntoutsi, Y. Theodoridis, and R. Schult. Monic: modeling and monitoring cluster transitions. In Tina Eliassi-Rad, Lyle H. Ungar, Mark Craven, and Dimitrios Gunopulos, editors, *Proceedings of the Twelfth ACM SIGKDD International Conference on Knowledge Discovery and Data Mining*, pages 706–711. ACM, 2006.

[287] D. Spitzner, S. J. Marron, and G. K. Essick. Mixed-model functional anova for studying human tactile perception. *Journal of the American Statistical Association*, 98:263–272, 2003.

[288] A. M. Staicu, C. M. Crainiceanu, and R. J. Carroll. Fast methods for spatially correlated multilevel functional data. *Biostatistics*, 11:177–194, 2010.

[289] R. Tang and H. G. Müller. Time-synchronized clustering of gene expression trajectories. *Biostatistics*, 10:850–858, 2009.

[290] M. A. Tanner and W. H. Wong. The calculation of posterior distributions by data augmentation. *Journal of the American Statistical Association*, 82(398):528–540, 1987.

[291] T. Tarpey, E. Petkova, and R. T. Ogden. Profiling placebo responders by self-consistent partitioning of functional data. *Journal of the American Statistical Association*, 98:850–858, 2003.

[292] R. Tibshirani. Regression shrinkage and selection via the Lasso. *Journal of the Royal Statistical Society Series B*, 58:267–288, 1996.

[293] E. F. Tjong Kim Sang and S. Buchholz. Introduction to the CoNLL-2000 shared task: Chunking. In C. Cardie, W. Daelemans, C. Nedellec, and E. F. Tjong Kim Sang, editors, *Proceedings of CoNLL-2000 and LLL-2000*, pages 127–132. Lisbon, Portugal, 2000.

[294] M. Tsatsos. *An Introduction to Topos Physics*. Master's thesis, University of London, 2008. http://arxiv.org/pdf/0803.2361v1.

[295] A. Tsybakov. Optimal aggregation of classifiers in statistical learning. *Annals of Statistics*, 32(1):135–166, 2004.

[296] D. A. van Dyk and X. L. Meng. The art of data augmentation. *Journal of Computational and Graphical Statistics*, 10(1):1–50, 2001.

[297] C. J. van Rijsbergen. *Information Retrieval*. Butterworths, London, second edition, 1979.

[298] C. J. van Rijsbergen. *The Geometry of Information Retrieval*. Cambridge University Press, 2004.

[299] A. Vanlierde, A. G. De Volder, M.-C. Wanet-Defalque, and C. Veraart. Occipito-parietal cortex activation during visuo-spatial imagery in early blind humans. *NeuroImage*, 19:698–709, 2003.

[300] V. Vapnik. *The Nature of Statistical Learning Theory*. Springer-Verlag, 1995.

[301] V. Vapnik. *Estimation of Dependencies Based on Empirical Data*. Springer, 2006. Reprint of 1982 Edition.

[302] V. Vapnik and A. Y. Chervonenkis. On the uniform convergence of relative frequencies of events to their probabilities. *Theory of Probability and its Applications*, 16(2):264–280, 1971.

[303] V. Vapnik and A. Y. Chervonenkis. Necessary and sufficient conditions for the uniform convergence of means to their expectations. *Theory of Probability and its Applications*, 26(3):532–553, 1981.

[304] V. Vapnik and A. Vashist. A new learning paradigm: learning using privileged information. *Neural Networks*, 22(5-6):544–557, 2009.

[305] V. Vapnik, A. Vashist, and N. Pavlovich. Learning using hidden information: master class learning. In *Proceedings of NATO workshop on Mining Massive Data Sets for Security*, pages 3–14. NATO ASI, 2008.

[306] R. Vignes. *Caractérisation automatique de groupes biologiques*. Thèse de doctorat, Université Paris VI, 1991.

[307] L. Von Ahn and L. Dabbish. Labeling images with a computer game. In *CHI*, pages 319–326, 2004.

[308] V. Vovk, A. Gammerman, and G. Shafer. *Algorithmic Learning in a Random World*. Springer, 2005.

[309] K. Wakita and T. Tsurumi. Finding community structure in mega-scale social networks. *Preprint*, 2007. http://arxiv.org/abs/cs.CY/0702048v1.

[310] S. Wang, W. Jank, and G. Shmueli. Explaining and forecasting online auction prices and their dynamics using functional data analysis. *Journal of Business Economics and Statistics*, 26(2):144–160, 2008.

[311] S. Wasserman and K. Faust. *Social Network Analysis: Methods and Applications.* Cambridge University Press, 1994.

[312] J. D. Watson and F. Crick. Molecular structure of nucleic acids: a structure for deoxyribose nucleic acid. *Nature*, 171:737–738, 1953.

[313] D. Watts. *Six Degrees, The Science of a Connected Age.* W.W. Norton, 2003.

[314] A. Webb. *Statistical Pattern Recognition.* Wiley, 2002.

[315] B. Widrow and M. E. Hoff. Adaptive switching circuits. In *IRE WESCON Conv. Record, Part 4*, pages 96–104, 1960.

[316] S. Wolfram. *A New Kind of Science.* Wolfram Media, 2002.

[317] W. Xu. Towards optimal one pass large scale learning with averaged stochastic gradient descent. Technical report, NEC Labs, 2010.

[318] F. Yao, H. G. Müller, and J. L. Wang. Functional linear regression analysis for longitudinal data. *Annals of Statistics*, 33(6):2873–2903, 2005.

[319] T. Yassine. *Aux origines d'une ethnosociologie singulière, in P. Bourdieu.* Seuil, 2008.

[320] T. Zhang, R. Ramakrishnan, and M. Livny. Birch: A new data clustering algorithm and its applications. *Data Mining and Knowledge Discovery*, 1:141–182, January 1997.

[321] X. H. Zhou, N.A. Obuchowski, and D.K. McClish. *Statistical Methods in Diagnostic Medicine.* Wiley, 2002.

# Index

## A
affinity, 176
AIC, 165
analysis
    document, 100
    semantic, 12, 92, 94, 96
    text, 92, 96
approximation error, 20
averaged stochastic gradient descent, 22

## B
background knowledge, 176
Bayesian analysis, 135
BIC, 165
biostatistics, 194
bootstrap, 127, 132
Brier score, 106, 168, 169

## C
c-index, 168
cancer mortality, 154
capital, 78, 85
    cultural, 78, 79
    economic, 79
    social, 79
    symbolic, 79
CART, 159
causal sets, 76
causets, 76
change detection, 115
chemometrics, 193
choriogenesis, 63, 64, 67, 68, 72, 73, 76
cinema, 91
class specific analysis, 88
classification threshold, 107–109, 111
classification trees, 159
classifier performance, 106
climatology, 194
cloud
    of individuals, 79, 81, 82, 87, 88
    of modalities, 83, 85, 86
cluster follow-up, 116
cluster ratio, 117

clustering, 114
    constrained, 93, 95, 98
    dynamic clustering, 171
    hierarchical, 92, 97, 98
co-clustering, 14–16
communities, 7, 16
conditional random field, 24
confidence, 44
conformal predictor, 45
correspondence, 67, 69, 72, 74
    between spaces, 73, 74
    symmetrical, 72, 74
correspondence analysis, 63, 66, 67, 72, 73, 76, 77, 92, 94, 97, 98, 100, 101
    between spaces, 73
    continuous, 68, 74
    discrete, 64, 72
    multiple, 77
cosmology, 64, 67, 76
credit card fraud, 13

## D
data
    augmentation, 135
    batch, 114
    imprecise, 161
    numeric, 161
    streams, 117
    symbolic, 161
data analysis
    functional, 189, 197
    social network, 3, 5, 9
    symbolic, 171, 172
data mining
    social network, 3, 5, 9
    text, 92, 96
descent algorithm, 158
deviance, 168
diabetes, 166
diagnosis, 103, 111
dimensionality reduction, 153
DNA, 125
document analysis, 100